T0073050

Class Groups of Number Fields and Related Topics

Kalyan Chakraborty · Azizul Hoque ·
Prem Prakash Pandey
Editors

Class Groups of Number Fields and Related Topics

Editors
Kalyan Chakraborty
School of Mathematics
Harish-Chandra Research Institute
Allahabad, Uttar Pradesh, India

Azizul Hoque
School of Mathematics
Harish-Chandra Research Institute
Allahabad, Uttar Pradesh, India

Prem Prakash Pandey
Department of Mathematics
IISER Berhampur
Berhampur, Odisha, India

ISBN 978-981-15-1516-3 ISBN 978-981-15-1514-9 (eBook)
https://doi.org/10.1007/978-981-15-1514-9

Mathematics Subject Classification (2010): 11Rxx, 11Sxx, 13C20

This Springer imprint is published by the registered company Springer Nature Singapore Pte Ltd.
The registered company address is: 152 Beach Road, #21-01/04 Gateway East, Singapore 189721, Singapore

To all concerned, who were responsible for the successful completion of ICCGNFRT-2017

Preface

The number theory seminar has been organized, from January 20, 2017, by Algebraic/Algorithmic/Analytic Number Theory Seminar (ANTS) at Harish-Chandra Research Institute, Allahabad, India. This lecture series was started by Kalyan Chakraborty, Azizul Hoque and other members of the group. Prior to the existence of this group, we had decided to hold a series of three conferences on the theme 'Class Groups of Number Fields and Related Topics.' By October 2019, we had organized these three conferences. However, seeing its success and also on the request of all concerned, we have decided to continue this yearly conference.

The first 'International Conference on Class Groups of Number Fields and Related Topics (ICCGNFRT)' was held during September 4–7, 2017, at Harish-Chandra Research Institute, Allahabad, India.

This collection comprises original research papers and survey articles presented at ICCGNFRT-2017. There are 16 chapters on important topics in algebraic number theory and related parts of analytic number theory. These topics include class groups and class numbers of number fields, units, the Kummer–Vandiver conjecture, class number one problem, Diophantine equations, Thue equations, continued fractions, Euclidean number fields, heights, rational torsion points on elliptic curves, cyclotomic numbers, Jacobi sums and Dedekind zeta values.

We are grateful to Springer and its mathematics editor(s), especially Mr. Shamim Ahmad, for publishing this volume.

Allahabad, India
October 2019

Kalyan Chakraborty
Azizul Hoque
Prem Prakash Pandey

Contents

A Geometric Approach to Large Class Groups: A Survey 1
Jean Gillibert and Aaron Levin

On Simultaneous Divisibility of the Class Numbers of Imaginary Quadratic Fields 17
Toru Komatsu

Thue Diophantine Equations 25
Michel Waldschmidt

A Lower Bound for the Class Number of Certain Real Quadratic Fields 43
Fuminori Kawamoto and Yasuhiro Kishi

A Survey of Certain Euclidean Number Fields 57
Kotyada Srinivas and Muthukrishnan Subramani

Divisibility of Class Number of a Real Cubic or Quadratic Field and Its Fundamental Unit 67
Anupam Saikia

The Charm of Units I, On the Kummer–Vandiver Conjecture. Extended Abstract 73
Preda Mihăilescu

Heights and Principal Ideals of Certain Cyclotomic Fields 89
René Schoof

Distribution of Residues Modulo *p* Using the Dirichlet's Class Number Formula 97
Jaitra Chattopadhyay, Bidisha Roy, Subha Sarkar and R. Thangadurai

On Class Number Divisibility of Number Fields and Points on Elliptic Curves 109
Debopam Chakraborty

Small Fields with Large Class Groups 113
Florian Luca and Preda Mihăilescu

Cyclotomic Numbers and Jacobi Sums: A Survey 119
Md. Helal Ahmed and Jagmohan Tanti

A Pair of Quadratic Fields with Class Number Divisible by 3 141
Himashree Kalita and Helen K. Saikia

On Lebesgue–Ramanujan–Nagell Type Equations 147
Richa Sharma

Partial Dedekind Zeta Values and Class Numbers of R–D Type Real Quadratic Fields 163
Mohit Mishra

On the Continued Fraction Expansions of $\sqrt{p}$ and $\sqrt{2p}$ for Primes $p \equiv 3 \pmod 4$ 175
Stéphane R. Louboutin

About the Editors

Kalyan Chakraborty is Professor at Harish-Chandra Research Institute (HRI), Allahabad, India, where he also obtained his Ph.D. in Mathematics. Professor Chakraborty was a postdoctoral fellow at IMSc, Chennai, and at Queen's University, Canada, and a visiting scholar at the University of Paris VI, VII, France; Tokyo Metropolitan University, Japan; Universitá Roma Tre, Italy; The University of Hong Kong, Hong Kong; Northwest University and Shandong University, China; Mahidol University, Thailand; Mandalay University, Myanmar; and many more. His broad area of research is number theory, particularly class groups, Diophantine equations, automorphic forms, arithmetic functions, elliptic curves, and special functions. He has published more than 60 research articles in respected journals and two books on number theory, and has been on the editorial boards of various leading journals. Professor Chakraborty is Vice-President of the Society for Special Functions and their Applications.

Azizul Hoque is a national postdoctoral fellow at Harish-Chandra Research Institute (HRI), Allahabad. He earned his Ph.D. in Pure Mathematics from Gauhati University, Guwahati, in 2015. Before joining HRI, Dr. Hoque was Assistant Professor at the Regional Institute of Science and Technology, Meghalaya, and at the University of Science and Technology, Meghalaya. He has visited Hong Kong University, Hong Kong; Northwest University, China; Shandong University, China; Mahidol University, Thailand; and many more. His research has mostly revolved around class groups, Diophantine equations, elliptic curves, zeta values, and related topics, and he has published a considerable number of papers in respected journals. He has been involved in a number of conferences and received numerous national and international grants.

Prem Prakash Pandey is Assistant Professor at the Indian Institute of Science Education and Research (IISER) Berhampur, Odisha. Before that, he was a post-doctoral fellow at HRI, Allahabad, and NISER Bhubaneswar, Odisha. After completing his Ph.D. at the Institute of Mathematical Sciences (IMSc), Chennai, he spent a couple of years at Chennai Mathematical Institute (CMI), Chennai, as a visiting scholar. Dr. Pandey's interests include class groups of number fields, annihilators of class groups, Diophantine equations, and related topics. During his time at HRI, he worked on divisibility problems for class numbers of quadratic fields with Dr. Hoque and Prof. Chakraborty.

A Geometric Approach to Large Class Groups: A Survey

Jean Gillibert and Aaron Levin

1 The Survey

The geometric techniques we shall report on are in fact explanations of geometric nature of a strategy that has been used from the beginning of the subject. We hope to convince the reader that this geometric viewpoint has many advantages. In particular, it clarifies the general strategy, and it allows one to obtain quantitative results. Furthermore, it raises new questions concerning torsion subgroups of Jacobians of curves defined over number fields.

Let us point out that, for simplicity, we focus here on geometric techniques related to covers of curves. Similar results hold for covers of arbitrary varieties, see [24] and [18], at the price of greater technicalities. The advantage of considering arbitrary varieties is that it provides a general framework to explain all constructions from previous authors, without exception.

1.1 Large Class Groups: The Folklore Conjecture

If M is a finite abelian group, and if $m > 1$ is an integer, we define the m-rank of M to be the maximal integer r such that $(\mathbb{Z}/m\mathbb{Z})^r$ is a subgroup of M; we denote it

The second author was supported in part by NSF grant DMS-1102563.

J. Gillibert (✉)
Institut de Mathématiques de Toulouse, CNRS UMR 5219,
118 route de Narbonne, 31062 Toulouse Cedex 9, France
e-mail: jean.gillibert@math.univ-toulouse.fr

A. Levin
Department of Mathematics, Michigan State University,
619 Red Cedar Road, East Lansing, MI 48824, USA
e-mail: adlevin@math.msu.edu

K. Chakraborty et al. (eds.), *Class Groups of Number Fields and Related Topics*,
https://doi.org/10.1007/978-981-15-1514-9_1

by $\mathrm{rk}_m\, M$. If k is a number field, we let $\mathrm{Cl}(k)$ denote the ideal class group of k, and $\mathrm{Disc}(k)$ denote the (absolute) discriminant of k.

The following conjecture is widely believed to be true.

Conjecture 1.1 *Let $d > 1$ and $m > 1$ be two integers. Then $\mathrm{rk}_m\,\mathrm{Cl}(k)$ is unbounded when k runs through the number fields of degree $[k : \mathbb{Q}] = d$.*

When $m = d$, and more generally when m divides d, this conjecture follows easily from Class Field Theory. On the other hand, when m and d are coprime, there is not a single case where Conjecture 1.1 is known to hold.

When constructing families of degree d fields k with a given lower bound on $\mathrm{rk}_m\,\mathrm{Cl}(k)$, it is natural to count the number of fields constructed, ordered by discriminant. This is the quantitative aspect of Conjecture 1.1.

For a detailed account of qualitative results towards Conjecture 1.1, and a discussion of quantitative results, see Sect. 1.4.

1.2 A Toy Example

In order to give a flavour of our technique, we revisit a classical construction.

Fact *Let $m \geq 3$ be an odd integer. For infinitely many odd $x \in \mathbb{N}$, the imaginary quadratic field $k = \mathbb{Q}(\sqrt{1 - x^m})$ satisfies $\mathrm{rk}_m\,\mathrm{Cl}(k) \geq 1$.*

Classical Proof

If $y = \sqrt{1 - x^m}$, then $y^2 = 1 - x^m$ and $x^m = (1 - y)(1 + y)$. The ideal $(1 - y, 1 + y)$ divides 2, but is also coprime to 2 because x is odd. Thus $1 - y$ and $1 + y$ are coprime, and their product is an mth power; hence each of them generates an ideal that is the mth power of some ideal of $\mathcal{O}_k$. So there exists an ideal $\mathfrak{a}$ of $\mathcal{O}_k$ such that $(1 - y) = \mathfrak{a}^m$. Therefore, the class of $\mathfrak{a}$ in $\mathrm{Cl}(k)$ has order dividing m.

The next (and hardest) step is to prove that, if some additional condition is satisfied, then the class of $\mathfrak{a}$ in $\mathrm{Cl}(k)$ has exact order m. Let us assume that the class of $\mathfrak{a}$ has order $q < m$, i.e. there exists $\alpha \in \mathcal{O}_k$ such that $\mathfrak{a}^q = (\alpha)$. One can write $m = q\ell$ for some odd $\ell > 1$. It follows that $(\alpha)^\ell = (1 - y)$, and hence there exists a unit ε of k such that $\varepsilon\alpha^\ell = 1 - y$. Assuming that the square-free part of $1 - x^m$ is strictly smaller than -3, the units of the imaginary quadratic field k are $\{\pm 1\}$, and hence $\alpha^\ell = 1 - y$ up to sign change. The result ultimately relies on the inexistence of solutions to this Diophantine equation, which one achieves by putting additional conditions on x. For example, Ram Murty [35] has shown that, if the square factor of $1 - x^m$ is less than $x^{m/4}/2\sqrt{2}$, then this equation has no solution. The same result is proved in [9] under the condition that x is a prime number ≥ 5. Eventually, as was pointed out by Cohn [13], it follows from the work of Nagell [30] that, for any odd number $x \geq 5$, one has $\mathrm{rk}_m\,\mathrm{Cl}(k) \geq 1$.

Quantitative results are usually obtained by *ad hoc* techniques of analytic number theory, depending on the required condition on x.

Geometric Proof

Let C be the smooth, projective, geometrically irreducible hyperelliptic curve over $\mathbb{Q}$ defined by the affine equation

$$y^2 = 1 - x^m.$$

Let $T \in C(\mathbb{Q})$ be the point with affine coordinates $(0, 1)$. The integer m being odd, the curve C has a unique point at infinity, that we denote by ∞. The divisor of the rational function $1 - y$ is given by

$$\operatorname{div}(1 - y) = mT - m\infty,$$

which proves that the class of the divisor $T - \infty$ defines a rational point of order dividing m in the Jacobian of C. It is not very hard to check that in fact, this divisor class has exact order m in the Jacobian of C.

A brief reminder on ramification in Kummer extensions : let K be a local field with valuation v, let $m > 1$ be an integer, and let $\gamma \in K^\times$. If the Kummer extension $K(\sqrt[m]{\gamma})/K$ is unramified at v, then $v(\gamma) \equiv 0 \pmod{m}$. Conversely, if $v(\gamma) \equiv 0 \pmod{m}$ and the residue characteristic of K is coprime to m, then $K(\sqrt[m]{\gamma})/K$ is unramified at v.

The valuation of $1 - y$ at each place of C is a multiple of m; hence the function field extension $\mathbb{Q}(C)(\sqrt[m]{1-y})/\mathbb{Q}(C)$ is unramified at each place of C. Therefore, this extension corresponds to an étale cover of C, that we denote by $f : \tilde{C} \to C$. This is a geometrically connected cover of degree m, because the class of $T - \infty$ has order m in the Jacobian of C.

Let us consider a point $P \in C(\overline{\mathbb{Q}})$ satisfying the following properties:

(i) for each finite place v of $\mathbb{Q}(P)$, $v(1 - y(P)) \equiv 0 \pmod{m}$;
(ii) $[\mathbb{Q}(\sqrt[m]{1 - y(P)}) : \mathbb{Q}(P)] = m$;
(iii) $\mathbb{Q}(P)$ is linearly disjoint from the mth cyclotomic field $\mathbb{Q}(\mu_m)$.

Then we claim that $\operatorname{rk}_m \operatorname{Cl}(\mathbb{Q}(P)) \geq 1 - \operatorname{rk}_{\mathbb{Z}} \mathcal{O}_{\mathbb{Q}(P)}^\times$. In order to prove this, let us define

$$\operatorname{Sel}^m(\mathbb{Q}(P)) := \{\gamma \in \mathbb{Q}(P)^\times/(\mathbb{Q}(P)^\times)^m;\ \forall v \text{ finite place of} \mathbb{Q}(P),\ v(\gamma) \equiv 0 \pmod{m}\}, \quad (1)$$

which is an analogue of the Selmer group for the multiplicative group over $\mathbb{Q}(P)$. Then we have an exact sequence

$$1 \longrightarrow \mathcal{O}_{\mathbb{Q}(P)}^\times/(\mathcal{O}_{\mathbb{Q}(P)}^\times)^m \longrightarrow \operatorname{Sel}^m(\mathbb{Q}(P)) \longrightarrow \operatorname{Cl}(\mathbb{Q}(P))[m] \longrightarrow 0.$$

By condition (i), the element $1 - y(P)$ defines a class in $\operatorname{Sel}^m(\mathbb{Q}(P))$, which has exact order m by condition (ii). It follows from condition (iii) that $\operatorname{rk}_m \mathcal{O}_{\mathbb{Q}(P)}^\times/(\mathcal{O}_{\mathbb{Q}(P)}^\times)^m =$

$\mathrm{rk}_{\mathbb{Z}}\, \mathcal{O}^{\times}_{\mathbb{Q}(P)}$. Therefore, by considering m-ranks in the exact sequence above, one obtains the result.

We shall now prove the existence of infinitely many points $P \in C(\overline{\mathbb{Q}})$ satisfying (i), (ii) and (iii), such that $\mathbb{Q}(P)$ is an imaginary quadratic field. It follows from Dirichlet's unit theorem that $\mathrm{rk}_m \mathrm{Cl}(\mathbb{Q}(P)) \geq 1$ for such fields. On the other hand, if $\mathbb{Q}(P)$ is a real quadratic field, then this machinery does not yield any result, because we do not have a way to ensure that $1 - y(P)$ is not a unit modulo mth powers.

It follows from an appropriate version of the Chevalley–Weil theorem that, if p is a prime of good reduction of C, then for any point $P \in C(\overline{\mathbb{Q}})$, the extension $\mathbb{Q}(\sqrt[m]{1 - y(P)})/\mathbb{Q}(P)$ is unramified at all places v dividing p. For such v, the condition $v(1 - y(P)) \equiv 0 \pmod{m}$ is satisfied, according to the Kummer criterion.

An immediate application of the Jacobian criterion of smoothness shows that the primes of bad reduction of C are the primes dividing $2m$. We shall now deal with local conditions at such primes p. Let P_0 be the rational point of C with affine coordinates $(1, 0)$. Then P_0 is a ramification point of x, and $y(P_0) = 0$. Let $P \in C(\overline{\mathbb{Q}})$ be a point which is p-adically close enough to P_0, by which we mean that, for each place v of $\mathbb{Q}(P)$ dividing p, the point P is close to P_0 for the v-adic topology on $C(\mathbb{Q}(P)_v)$. Then by elementary considerations $y(P)$ is p-adically close to $y(P_0)$, hence to 0. Therefore, if P is p-adically close enough to P_0, then $v(1 - y(P)) = 0$ for each place v of $\mathbb{Q}(P)$ dividing p.

Let Δ be the product of bad primes, and let $\phi : C \to \mathbb{P}^1$ be the rational map defined by $\phi = \frac{x-1}{\Delta^N}$ for some integer N large enough. The map ϕ being totally ramified at P_0, one can see that, for all $t \in \mathbb{N}$ and all bad primes p, the point $P_t := \phi^{-1}(t)$ is p-adically close enough to the point P_0. Then the discussion above proves that all points P_t with $t \in \mathbb{N}$ satisfy condition (i). In fact, it was shown in the classical proof that the map $\phi = \frac{x-1}{2}$ does the job, so we shall use that one instead.

Applying Hilbert's irreducibility theorem to the composite cover of degree $2m$

$$\tilde{C} \xrightarrow{\ f\ } C \xrightarrow{\ \phi\ } \mathbb{P}^1,$$

we obtain the existence of infinitely many $t \in \mathbb{N}$ such that $[\mathbb{Q}(f^{-1}(P_t)) : \mathbb{Q}] = 2m$. For such t, the field $\mathbb{Q}(P_t)$ is quadratic, and $\mathbb{Q}(f^{-1}(P_t)) = \mathbb{Q}(\sqrt[m]{1 - y(P_t)})$ is an extension of degree m of $\mathbb{Q}(P_t)$. Hence condition (ii) is satisfied. Moreover, $\mathbb{Q}(P_t) = \mathbb{Q}(\sqrt{1 - (2t + 1)^m})$ is imaginary quadratic, hence (iii) holds (unless $3 \mid m$ and $\mathbb{Q}(P) = \mathbb{Q}(\sqrt{-3})$, which we exclude). This concludes the proof of the statement.

Finally, it follows from a quantitative version of Hilbert's irreducibility theorem, due to Dvornicich and Zannier [17], that, given $X > 0$, there exist $\gg X^{\frac{1}{m}} / \log X$ imaginary quadratic fields $\mathbb{Q}(P_t)$ with discriminant $|\,\mathrm{Disc}(\mathbb{Q}(P_t))| < X$ such that condition (ii) is satisfied. This yields a quantitative version of the result.

Comments

At first glance, the geometric proof seems more technical than the classical one. Let us list some advantages of this technique over the classical one.

The first advantage of geometry is to avoid the use of *ad hoc* "tricks". More precisely, in the classical method one uses two tricks: the first one is to ensure that

$1 - y$ is the mth power of some ideal of k, which is done by requiring congruence conditions on the variables. In the geometric case, this relies on the Chevalley–Weil theorem and some additional condition, namely there exists a rational point on C at which the map $C \to \mathbb{P}^1$ is totally ramified. The second trick, which is the hardest, is to prove that, for every $\ell > 1$ dividing m, the equation $\alpha^\ell = 1 - y$ has no solution $\alpha \in k$. In the geometric world, this follows immediately from Hilbert's irreducibility theorem, without any additional technicality.

Another nice feature of the geometric approach: the use of Hilbert's irreducibility theorem automatically gives us a quantitative version of the result. This should be compared to the specific analytic number theory machinery that has been used previously on these quantitative class group problems.

A final advantage comes from the (arguable) fact that it is relatively easier to build étale covers of curves than everywhere unramified extensions of number fields.

1.3 General Specialization Results

Following the lines of the geometric proof of the "toy example" above, one obtains the following general statement.

Theorem 1.2 *Let C be a smooth, projective, geometrically irreducible curve over $\mathbb{Q}$, let $\mathrm{Jac}(C)$ be the Jacobian of C, and let $m > 1$ be an integer. Assume that C admits a finite morphism $C \to \mathbb{P}^1$ of degree d, totally ramified over some point belonging to $\mathbb{P}^1(\mathbb{Q})$. Then there exist infinitely many (isomorphism classes of) number fields k with $[k : \mathbb{Q}] = d$ such that*

$$\mathrm{rk}_m \, \mathrm{Cl}(k) \geq \mathrm{rk}_m \, \mathrm{Jac}(C)(\mathbb{Q})_{\mathrm{tors}} - \mathrm{rk}_{\mathbb{Z}} \, \mathcal{O}_k^\times . \tag{2}$$

Inspired by the technique introduced in [24], this theorem was proved in [18] in the case when C is a superelliptic curve defined by a "nice equation" (see Corollary 3.1 of [18]). The version above is proved in [3], the base field being $\mathbb{Q}$ for simplicity. In Sect. 2.4 we state a variant of this result, in which the assumption that the morphism $C \to \mathbb{P}^1$ is totally ramified over some rational point is replaced by a more technical one.

While Theorem 1.2 is quite general, its applicability in concrete cases is impaired by the presence of the negative term $-\mathrm{rk}_{\mathbb{Z}} \, \mathcal{O}_k^\times$ on the right: the rank of the unit group of the field k tends to be large, especially if $d > 2$.

This deficiency is avoided in the following theorem, which constitutes the main result of [3]. Let us denote by $\mathrm{rk}_{\mu_m} \mathrm{Jac}(C)$ the maximal integer r such that $\mathrm{Jac}(C)$ has a $\mathrm{Gal}(\overline{\mathbb{Q}}/\mathbb{Q})$-submodule isomorphic to μ_m^r.

Theorem 1.3 *In the setup of Theorem 1.2, there exist infinitely many number fields k with $[k : \mathbb{Q}] = d$ such that*

$$\mathrm{rk}_m \, \mathrm{Cl}(k) \geq \mathrm{rk}_{\mu_m} \mathrm{Jac}(C). \tag{3}$$

In contradistinction with the proof of Theorem 1.2 which is based on Kummer theory, the proof of Theorem 1.3 relies on Class Field Theory. It can be seen as a generalization of the constructions of Mestre [25–28].

In both Theorems 1.2 and 1.3 the "infinitely many" can be made quantitative, the fields being ordered by discriminant.

Theorem 1.4 *Let $\phi \in \mathbb{Q}(C)$ be the rational function defining the morphism $C \to \mathbb{P}^1$ appearing in both Theorems 1.2 and 1.3. Assume that there exists a rational function $x \in \mathbb{Q}(C)$ of degree n such that $\mathbb{Q}(C) = \mathbb{Q}(\phi, x)$. Then, for sufficiently large positive X, in both these theorems the number of isomorphism classes of the fields k satisfying* (2) *or* (3)*, respectively, and such that $|\operatorname{Disc}(k)| \leq X$ is $\gg X^{1/2n(d-1)}/\log X$.*

In a recent work [5], Bilu and Luca improved the quantitative version of Hilbert's irreducibility theorem given by Dvornicich and Zannier, on which our quantitative results are based. We underline the fact that any improvement of quantitative HIT automatically yields a similar improvement in our quantitative results.

In the case when C is a hyperelliptic curve with a rational Weierstrass point, it is possible to improve slightly the quantitative result. More precisely, we obtain in [18] the following quantitative version of Theorem 1.2 for such curves.

Corollary 1.5 *Let C be a smooth projective hyperelliptic curve over $\mathbb{Q}$ with a rational Weierstrass point, and let $m > 1$ be an integer. Let g denote the genus of C. Then there exist $\gg X^{\frac{1}{2g+1}}/\log X$ imaginary (resp. real) quadratic number fields k with $|\operatorname{Disc}(k)| < X$ and*

$$\operatorname{rk}_m \operatorname{Cl}(k) \geq \operatorname{rk}_m \operatorname{Jac}(C)(\mathbb{Q})_{\mathrm{tors}}$$
$$(\textit{resp.}\ \operatorname{rk}_m \operatorname{Cl}(k) \geq \operatorname{rk}_m \operatorname{Jac}(C)(\mathbb{Q})_{\mathrm{tors}} - 1).$$

In view of the statement above, the following question arises immediately.

Question 1.6 *Let $m > 1$ be an integer. Do there exist hyperelliptic curves C over $\mathbb{Q}$ with $\operatorname{rk}_m \operatorname{Jac}(C)(\mathbb{Q})_{\mathrm{tors}}$ arbitrarily large?*

According to Corollary 1.5, a positive answer to this question would provide a proof of Conjecture 1.1 in the $d = 2$ case, provided the curves have rational Weierstrass points.

For $m = 2$, the question above has a positive answer. Apart from this easy case, very little is known. To our knowledge, the best general result is the following: given $m > 1$, there exist hyperelliptic curves C over $\mathbb{Q}$ with $\operatorname{rk}_m \operatorname{Jac}(C)(\mathbb{Q})_{\mathrm{tors}} \geq 2$. This allows one to derive Yamamoto's result from Corollary 1.5 (see Sect. 2.1).

Finally, it follows from Theorem 1.3 that Corollary 1.5 and Question 1.6 have natural analogues in which $\operatorname{rk}_m \operatorname{Jac}(C)(\mathbb{Q})_{\mathrm{tors}}$ is replaced by $\operatorname{rk}_{\mu_m} \operatorname{Jac}(C)$. Unfortunately, we have not been able to find examples of hyperelliptic curves over $\mathbb{Q}$ with large $\operatorname{rk}_{\mu_m} \operatorname{Jac}(C)$.

1.4 *Record of Known Results Towards Conjecture 1.1*

In this section, we give a brief summary of the history of results on the problem of finding infinite families of number fields of degree d over $\mathbb{Q}$ with ideal class groups of large m-rank (see Tables 1 and 2 for a more comprehensive list of results). The earliest such result could be considered to be Gauss' result determining, in modern terms, the 2-rank of the class group of a quadratic number field in terms of the primes dividing the discriminant of the quadratic field. In particular, it follows from Gauss' result that the 2-rank of the ideal class group of a quadratic number field can be made arbitrarily large. In contrast to Gauss' result, there is not a single quadratic number field k and prime $p \neq 2$ for which it is known that $\text{rk}_p \text{Cl}(k) > 6$, although the Cohen–Lenstra heuristics [11, 12] predict that for any given positive integer r, a positive proportion of quadratic fields k should have $\text{rk}_p \text{Cl}(k) = r$.

The first constructive result on m-ranks of class groups for arbitrary m was given in 1922 by Nagell [29, 31], who proved that for any positive integer m, there exist infinitely many imaginary quadratic number fields whose class group has an element of order m (in particular, there are infinitely many imaginary quadratic fields with class number divisible by m). Nagell's result has since been reproved by a number of different authors (e.g. [1, 20, 23]). Nearly 50 years later, working independently, Yamamoto [40] and Weinberger [39] extended Nagell's result to real quadratic fields. Soon after, Uchida [38] proved the analogous result for cubic cyclic fields. In 1984, Azuhata and Ichimura [2] succeeded in extending Nagell's result to number fields of arbitrary degree. In fact, they proved that for any integers $m, d > 1$ and any

Table 1 Values of m and r for which it is known that there exist infinitely many quadratic fields k with $\text{rk}_m \text{Cl}(k) \geq r$ (we let $r = \infty$ if $\text{rk}_m \text{Cl}(k)$ can be made arbitrarily large). All results in this table can be recovered by applying Corollary 1.5, except for Mestre's ones, which are applications of Theorem 1.3

Author(s)	Year	Type	m	r
Gauss	19th century	Imaginary, real	2	∞
Nagell [29, 31]	1922	Imaginary	> 1	1
Yamamoto [40]	1970	Imaginary	> 1	2
Yamamoto [40], Weinberger [39]	1970, 1973	Real	> 1	1
Craig [14]	1973	Imaginary	3	3
		Real	3	2
Craig [15]	1977	Imaginary	3	4
		Real	3	3
Diaz y Diaz [16]	1978	Real	3	4
Mestre [25–27]	1980	Imaginary, real	5, 7	2
Mestre [28]	1992	Imaginary, real	5	3

Table 2 Values of m, d and r for which it is known that there exist infinitely many number fields k of degree d with $\mathrm{rk}_m\,\mathrm{Cl}(k) \geq r$. All results in this table can be recovered by applying variants of Theorem 1.2, except for the cases when $m = 2$, which follow from variants of Theorem 1.3

Author(s)	Year	m	d	r
Brumer, Brumer and Rosen [6, 7]	1965	>1	$d = m$	∞
Uchida [38]	1974	>1	3	1
Ishida [21]	1975	2	Prime	$d - 1$
Azuhata and Ichimura [2]	1984	>1	>1	$\lfloor \frac{d}{2} \rfloor$
Nakano [32, 33]	1984	>1	>1	$\lfloor \frac{d}{2} \rfloor + 1$
	1985	2	>1	d
Nakano [34]	1988	2	3	6
Levin [24]	2007	>1	>1	$\lceil \lfloor \frac{d+1}{2} \rfloor + \frac{d}{m-1} - m \rceil$
Kulkarni [22]	2017	2	3	8

nonnegative integers r_1, r_2, with $r_1 + 2r_2 = d$, there exist infinitely many number fields k of degree $d = [k : \mathbb{Q}]$ with r_1 real places and r_2 complex places such that

$$\mathrm{rk}_m\,\mathrm{Cl}(k) \geq r_2. \tag{4}$$

The right-hand side of (4) was subsequently improved to $r_2 + 1$ by Nakano [32, 33]. Choosing r_2 as large as possible, we thus obtain, for any m, infinitely many number fields k of degree $d > 1$ with

$$\mathrm{rk}_m\,\mathrm{Cl}(k) \geq \left\lfloor \frac{d}{2} \right\rfloor + 1, \tag{5}$$

where $\lfloor \cdot \rfloor$ and $\lceil \cdot \rceil$ denote the greatest and least integer functions, respectively. For general m and d, (5) is the best result that is known on producing number fields of degree d with a class group of large m-rank. In [24], it was shown that there exist infinitely many number fields k of degree d satisfying $\mathrm{rk}_m\,\mathrm{Cl}(k) \geq \lceil \lfloor \frac{d+1}{2} \rfloor + \frac{d}{m-1} - m \rceil$, improving (5) when $d \geq m^2$.

For certain special values of m and d, slightly more is known. Of particular note to us are Mestre's papers [25–28] giving the best known results for $m = 5, 7$ and $d = 2$. Mestre's method can be seen as an application of Theorem 1.3 (see Sect. 2.3).

Recently, progress has been made on obtaining quantitative results on counting the number fields in the above results. Murty [36] gave the first results in this direction, obtaining quantitative versions of the theorems of Nagell and Yamamoto–Weinberger. His results have since been improved by, among others, Soundararajan [37] in the imaginary quadratic case and Yu [41] in the real quadratic case. In higher degrees, Hernández and Luca [19] gave the first such result for cubic number fields, while Bilu and Luca [4] succeeded in proving a quantitative theorem for number fields of arbitrary degree. Bilu and Luca's result was improved in [24], where a quantitative version of Azuhata and Ichimura's result was given. In Sect. 2.4, we show how it is possible to derive from Theorem 1.2 a short proof of this result.

2 The Examples

This section is devoted to examples of applications of Theorems 1.2 and 1.3. Each of these examples is obtained by revisiting previous constructions. In certain cases, this yields new quantitative results.

2.1 *Yamamoto's Result*

In [40], Yamamoto proved that, for any integer $m > 1$, there exist infinitely many imaginary (resp. real) quadratic fields k with $\mathrm{rk}_m \mathrm{Cl}(k) \geq 2$ (resp. $\mathrm{rk}_m \mathrm{Cl}(k) \geq 1$).

In order to recover this result via geometry, we proved the following in [18].

Lemma 2.1 *Let $\lambda \in \mathbb{Q}^\times$, $\lambda \neq \pm 1$, and let $m > 1$ be an integer. Let C be the smooth projective hyperelliptic curve defined over $\mathbb{Q}$ by the affine equation $y^2 = x^{2m} - (1 + \lambda^2)x^m + \lambda^2$. Then C has a rational Weierstrass point, and* $\mathrm{rk}_m \mathrm{Jac}(C)(\mathbb{Q})_{\mathrm{tors}} \geq 2$.

Applying Corollary 1.5 to this situation, we obtained the following quantitative version of Yamamoto's result [18, Corollary 3.4].

Corollary 2.2 *Let $m > 1$ be an integer. There exist $\gg X^{\frac{1}{2m-1}} / \log X$ imaginary (resp. real) quadratic number fields k with $|\mathrm{Disc}(k)| < X$ and* $\mathrm{rk}_m \mathrm{Cl}(k) \geq 2$ *(resp.* $\mathrm{rk}_m \mathrm{Cl}(k) \geq 1$*).*

If m is odd, then Byeon [8] and Yu [41] have proved, for imaginary and real quadratic fields, respectively, the better lower bound of $\gg X^{1/m-\epsilon}$. If m is even, in the real quadratic case a lower bound of $\gg X^{1/m}$ was proved by Chakraborty, Luca and Mukhopadhyay [10]. The imaginary quadratic case of Corollary 2.2 with m even appears to be a new result of [18].

2.2 *3-Ranks of Quadratic Fields: A Construction of Craig*

In [15], Craig constructed infinitely many imaginary (resp. real) quadratic fields k with $\mathrm{rk}_3 \mathrm{Cl}(k) \geq 4$ (resp. with $\mathrm{rk}_3 \mathrm{Cl}(k) \geq 3$). We prove quantitative versions of Craig's result and show how his constructions yield a hyperelliptic curve whose Jacobian has a rational subgroup isomorphic to $(\mathbb{Z}/3\mathbb{Z})^4$.

Let f be the polynomial

$$f(x, y, z) = x^6 + y^6 + z^6 - 2x^3y^3 - 2x^3z^3 - 2y^3z^3.$$

The idea in [15] is to find a nontrivial parametric family of solutions to the equations

$$f(x_0, y_0, z_0) = f(x_1, y_1, z_1) = f(x_2, y_2, z_2).$$

Since

$$f(x, y, z) = (x^3 + y^3 - z^3)^2 - 4x^3y^3 = (x^3 - y^3 + z^3)^2 - 4x^3z^3 = (-x^3 + y^3 + z^3)^2 - 4y^3z^3,$$

it suffices to find solutions to

$$x_1z_1 = x_0z_0, \quad x_2y_2 = x_0y_0, \tag{6}$$
$$x_1^3 - y_1^3 + z_1^3 = -(x_0^3 - y_0^3 + z_0^3), \tag{7}$$
$$x_2^3 + y_2^3 - z_2^3 = -(x_0^3 + y_0^3 - z_0^3). \tag{8}$$

Craig gives a two-parameter family of solutions to (6), (7) and (8) in terms of α, β and γ satisfying $\alpha + \beta + \gamma = 0$. We refer the reader to [15] for the rather involved formulas. We specialize Craig's solution by setting $\alpha = 0$, $\beta = t$ and $\gamma = -t$. This gives a polynomial $h(t) = f(x_0(t), y_0(t), z_0(t))$ of degree 141. Let C be the (non-singular projective model of the) hyperelliptic curve defined by $Y^2 = h(t)$. We have the four identities (where $x_0 = x_0(t)$, $y_0 = y_0(t)$, etc.),

$$\left(Y + (x_0^3 + y_0^3 - z_0^3)\right)\left(Y - (x_0^3 + y_0^3 - z_0^3)\right) = -4x_0^3y_0^3,$$
$$\left(Y + (x_0^3 - y_0^3 + z_0^3)\right)\left(Y - (x_0^3 - y_0^3 + z_0^3)\right) = -4x_0^3z_0^3,$$
$$\left(Y + (x_1^3 + y_1^3 - z_1^3)\right)\left(Y - (x_1^3 + y_1^3 - z_1^3)\right) = -4x_1^3y_1^3,$$
$$\left(Y + (-x_2^3 + y_2^3 + z_2^3)\right)\left(Y - (-x_2^3 + y_2^3 + z_2^3)\right) = -4y_2^3z_2^3.$$

It follows that there are divisors D_1, D_2, D_3 and D_4 on C such that

$$(Y + x_0^3 + y_0^3 - z_0^3) = 3D_1,$$
$$(Y + x_0^3 - y_0^3 + z_0^3) = 3D_2,$$
$$(Y + x_1^3 + y_1^3 - z_1^3) = 3D_3,$$
$$(Y - x_2^3 + y_2^3 + z_2^3) = 3D_4.$$

Using Magma, it is easy to verify that D_1, D_2, D_3 and D_4 give independent 3-torsion elements of $\mathrm{Jac}(C)(\mathbb{Q})$ (to simplify calculations, this can be done modulo $p = 7$, a prime of good reduction of C). Thus, we arrive at the following result.

Theorem 2.3 *Let C be the hyperelliptic curve defined by $Y^2 = h(t)$. Then*

$$\mathrm{rk}_3 \operatorname{Jac}(C)(\mathbb{Q})_{\mathrm{tors}} \geq 4.$$

Since h has odd degree, C has a rational Weierstrass point, and so Corollary 1.5 applies.

Corollary 2.4 *There exist $\gg X^{\frac{1}{141}} / \log X$ imaginary (resp. real) quadratic fields k with $|\operatorname{Disc}(k)| < X$ and $\mathrm{rk}_3 \operatorname{Cl}(k) \geq 4$ (resp. $\mathrm{rk}_3 \operatorname{Cl}(k) \geq 3$).*

2.3 5-Ranks of Quadratic Fields: A Construction of Mestre

In [28], Mestre proved the existence of infinitely many imaginary and real quadratic fields k with $\mathrm{rk}_5 \operatorname{Cl}(k) \geq 3$. We briefly review his construction. For the reader's convenience, we stick to the original notation. Mestre constructs

(1) a genus 5-hyperelliptic curve C defined over $\mathbb{Q}$, which admits three rational Weierstrass points;
(2) three elliptic curves E_1, E_2 and E_3 defined over $\mathbb{Q}$, each of them endowed with an isogeny $\varphi_i : E_i \to F_i$ with kernel $\mathbb{Z}/5\mathbb{Z}$;
(3) three independent Galois covers $\tau_i : C \to F_i$ with group $(\mathbb{Z}/2\mathbb{Z})^2$.

The existence of the maps τ_i implies that the Jacobian of C splits, and that each of the F_i is an isogenus factor of $\operatorname{Jac}(C)$ via an isogeny of degree 4. More precisely, there exists an abelian surface B and an isogeny

$$F_1 \times F_2 \times F_3 \times B \longrightarrow \operatorname{Jac}(C)$$

whose degree is a power of 2.

On the other hand, the dual isogeny $\hat{\varphi}_i : F_i \to E_i$ has kernel μ_5, because the kernel of the $\hat{\varphi}_i$ is the Cartier dual of the kernel of φ_i. Hence $\operatorname{Jac}(C)$ contains μ_5^3 as a subgroup, which means in our terminology that $\mathrm{rk}_{\mu_5} \operatorname{Jac}(C) \geq 3$.

Applying Theorem 1.3 to this situation, we obtain the following quantitative version of Mestre's result.

Theorem 2.5 *There exist $\gg X^{\frac{1}{11}} / \log X$ imaginary (resp. real) quadratic fields k with $|\operatorname{Disc}(k)| < X$ such that $\mathrm{rk}_5 \operatorname{Cl}(k) \geq 3$.*

2.4 Higher Degree Fields

Let us fix integers $m, r > 1$ with $(r, m) = 1$. Consider a superelliptic curve C defined by an affine equation of the form

$$y^m = a_0 \prod_{i=1}^{r} (x - a_i),$$

where $a_1, \ldots, a_r$ are pairwise distinct rational numbers, and $a_0 \in \mathbb{Q}^\times$. Then x and y are rational functions on C with $\deg x = m$ and $\deg y = r$. Since $(r, m) = 1$, the curve C has a unique point at infinity, that we denote by ∞, and x and y are totally ramified at that point. For each i, let P_i be the rational point on C with affine coordinates $(a_i, 0)$. Then one has

$$\operatorname{div}(x - a_i) = mP_i - m\infty.$$

A classical argument shows that the divisor classes $(P_i - \infty)_{i=1}^r$ generate a subgroup of $\operatorname{Jac}(C)(\mathbb{Q})$ isomorphic to $(\mathbb{Z}/m\mathbb{Z})^{r-1}$.

Applying Theorem 1.2 to the map $x : C \to \mathbb{P}^1$, one recovers the result of Brumer and Rosen (first line in Table 2). By considering the map $y : C \to \mathbb{P}^1$, one recovers results of Azuhata and Ichimura (line 4 in Table 2). Using Hilbert's irreducibility theorem, quantitative versions of these results were obtained in [24].

Using other maps, it was shown in [24] that in some situations it is possible to improve on Nakano's inequality (5) (line 5 in Table 2).

Theorem 2.6 *Let $m, d > 1$ be integers with $d > (m-1)^2$. There exist $\gg X^{\frac{1}{(m+1)d-1}} / \log X$ number fields k of degree d with $|\operatorname{Disc}(k)| < X$ and*

$$\operatorname{rk}_m \operatorname{Cl}(k) \geq \left\lceil \left\lfloor \frac{d+1}{2} \right\rfloor + \frac{d}{m-1} - m \right\rceil.$$

A detailed proof of this theorem is given in [24], but it is possible to give a simpler proof by using the following variant of Theorem 1.2.

Theorem 2.7 *Let C be a smooth projective geometrically irreducible curve over $\mathbb{Q}$, let $\operatorname{Jac}(C)$ be the Jacobian of C, and let $m > 1$ be an integer. Let $s = \operatorname{rk}_m \operatorname{Jac}(C)(\mathbb{Q})_{\text{tors}}$, and let $D_1, \ldots, D_s$ be divisors on C whose classes in $\operatorname{Jac}(C)(\mathbb{Q})$ generate a subgroup isomorphic to $(\mathbb{Z}/m\mathbb{Z})^s$. Let $g_1, \ldots, g_s$ be rational functions on C such that $\operatorname{div}(g_i) = mD_i$ for all i. Assume that there exists a finite map $\phi : C \to \mathbb{P}^1$ of degree d such that, for all $t \in \mathbb{N}$, the point $P_t := \phi^{-1}(t)$ has the property that*

$$g_1(P_t), \ldots, g_s(P_t) \text{ define classes in } \operatorname{Sel}^m(\mathbb{Q}(P_t)), \tag{9}$$

where Sel^m is defined in (1). Then there exist infinitely many $t \in \mathbb{N}$ such that $[\mathbb{Q}(P_t) : \mathbb{Q}] = d$ and

$$\operatorname{rk}_m \operatorname{Cl}(\mathbb{Q}(P_t)) \geq s - \operatorname{rk}_{\mathbb{Z}} \mathcal{O}_{\mathbb{Q}(P_t)}^\times.$$

Moreover, there are infinitely many isomorphism classes of such fields $\mathbb{Q}(P_t)$.

This statement is a generalization of Theorem 1.2, in which the condition on the existence of a totally ramified point for the map ϕ is replaced by a more technical one. In explicit examples, this technical condition usually comes from a congruence condition on the coordinates of the point P_t (see proof of Theorem 2.6).

Theorem 2.7 can be proved along the lines of the toy example. We refer to the proof of [18, Theorem 2.4], in which all needed arguments already appear. Needless to say, Theorem 2.7 admits the same quantitative version as the previous ones, as stated in Theorem 1.4.

Proof (*Proof of Theorem* 2.6) Let $m, d > 1$ be integers with $d > (m-1)^2$. Let r be the largest integer such that $r - \lfloor \frac{r}{m} \rfloor \leq d$ and $(r, m) = 1$. It is easily checked that $r \geq d + \frac{d}{m-1} - m + 1$. Let C be the curve defined by

$$y^m = h(x) = -(x - a_1^m) \prod_{i=2}^{r} (x + a_i^m),$$

where $a_1, \ldots, a_r$ are certain carefully chosen integers [24, Lemma 3.1]. For $i = 2, \ldots, r$, we let $g_i := x + a_i^m$. Then $\mathrm{div}(g_i) = mP_i - m\infty$ where $P_i = (-a_i^m, 0)$, and, as noted above, the divisor classes $(P_i - \infty)_{i=2}^{r}$ generate a subgroup isomorphic to $(\mathbb{Z}/m\mathbb{Z})^{r-1}$ in $\mathrm{Jac}(C)(\mathbb{Q})$.

Let $f(x)$ be the Taylor series for $\sqrt[m]{h(x)}$ at $x = 0$ truncated to degree $\lfloor \frac{r}{m} \rfloor - 1$ with $f(0) = \prod_{i=1}^{r} a_i$. Then f is defined over $\mathbb{Q}$, and

$$\mathrm{ord}_x(f^m - h) \geq \left\lfloor \frac{r}{m} \right\rfloor \geq r - d.$$

Let b be the lowest common denominator of the coefficients of f. Let $\psi : C \to \mathbb{P}^1$ be the rational function defined by

$$\psi := \frac{b(y - f)}{x^{r-d}}.$$

Then one computes [24, Lemma 3.5] that ψ has degree d.

Let Δ_0 be the product of prime numbers dividing the discriminant of h. Having chosen $a_1, \ldots, a_r$ properly, it can be shown [24, Lemma 3.3] that there exists an integer c_0 such that, for each integer $c \equiv c_0 \pmod{\Delta_0}$, the point $Q_c := \psi^{-1}(c)$ has the property that $g_2(Q_c), \ldots, g_r(Q_c)$ define classes in $\mathrm{Sel}^m(\mathbb{Q}(Q_c))$.

If we define $\phi : C \to \mathbb{P}^1$ by

$$\phi := \frac{\psi - c_0}{\Delta_0},$$

then ϕ also has degree d and, for all $t \in \mathbb{N}$, the point $P_t := \phi^{-1}(t)$ satisfies condition (9) from Theorem 2.7 with respect to the functions $g_2, \ldots, g_r$.

Finally, it follows from [24, Lemma 3.4] that

$$\mathrm{Disc}(\mathbb{Q}(P_t)) = O(t^{(m+1)d-1})$$

and $\mathbb{Q}(P_t)$ has at most two real places for $t \gg 0$. The result follows from Theorem 2.7. □

Acknowledgements The first author warmly thanks Yuri Bilu for very inspiring conversations, and for his feedback on a preliminary version of this note.

References

1. N.C. Ankeny, S. Chowla, On the divisibility of the class number of quadratic fields. Pac. J. Math. **5**(3), 321–324 (1955)
2. T. Azuhata, H. Ichimura, On the divisibility problem of the class numbers of algebraic number fields. J. Fac. Sci. Univ. Tokyo Sect. IA Math. **30**(3), 579–585 (1984)
3. Y.F. Bilu, J. Gillibert, Chevalley-Weil theorem and subgroups of class groups. Isr. J. Math. **226**(2), 927–956 (2018)
4. Y.F. Bilu, F. Luca, Divisibility of class numbers: enumerative approach. J. Reine Angew. Math. **578**, 79–91 (2005)
5. Y.F. Bilu, F. Luca, Diversity in parametric families of number fields, in *Number Theory-Diophantine Problems, Uniform Distribution and Applications* (Springer, Cham, 2017), pp. 169–191
6. A. Brumer, Ramification and class towers of number fields. Mich. Math. J. **12**(2), 129–131 (1965)
7. A. Brumer, M. Rosen, Class number and ramification in number fields. Nagoya Math. J. **23**, 97–101 (1963)
8. D. Byeon, Imaginary quadratic fields with noncyclic ideal class groups. Ramanujan J. **11**(2), 159–163 (2006)
9. K. Chakraborty, A. Hoque, Y. Kishi, P.P. Pandey, Divisibility of the class numbers of imaginary quadratic fields. J. Number Theory **185**, 339–348 (2018)
10. K. Chakraborty, F. Luca, A. Mukhopadhyay, Exponents of class groups of real quadratic fields. Int. J. Number Theory **4**(4), 597–611 (2008)
11. H. Cohen, H.W. Lenstra Jr., Heuristics on class groups, in *Number theory*. Lecture Notes in Mathematics, vol. 1052 (Springer, Berlin, 1984), pp. 26–36
12. H. Cohen, H.W. Lenstra Jr., Heuristics on class groups of number fields, in Number Theory. Lecture Notes in Mathematics, vol. 1068 (Springer, Berlin, 1984), pp. 33–62
13. J.H.E. Cohn, On the Diophantine equation $x^n = Dy^2 + 1$. Acta Arith. **106**, 73–83 (2003)
14. M. Craig, A type of class group for imaginary quadratic fields. Acta Arith. **22**, 449–459 (1973)
15. M. Craig, A construction for irregular discriminants. Osaka J. Math. **14**(2), 365–402 (1977)
16. F. Diaz y Diaz, *Sur le 3-rang des corps quadratiques*, Publications Mathématiques d'Orsay 78, vol. 11 (Université de Paris-Sud, Département de Mathématique, Orsay, 1978)
17. R. Dvornicich, U. Zannier, Fields containing values of algebraic functions. Ann. Scuola Norm. Sup. Pisa Cl. Sci. (4) **21**(3), 421–443 (1994)
18. J. Gillibert, A. Levin, Pulling back torsion line bundles to ideal classes. Math. Res. Lett. **19**(05), 1–14 (2012)
19. S. Hernández, F. Luca, Divisibility of exponents of class groups of pure cubic number fields, High primes and misdemeanours: lectures in honour of the 60th birthday of Hugh Cowie Williams. Fields Inst. Commun. **41**, 237–244 (2004). Amer. Math. Soc., Providence, RI

20. P. Humbert, Sur les nombres de classes de certains corps quadratiques. Comment. Math. Helv. **12**(1), 233–245 (1940)
21. M. Ishida, On 2-rank of the ideal class groups of algebraic number fields. J. Reine Angew. Math. **273**, 165–169 (1975)
22. A. Kulkarni, An explicit family of cubic number fields with large 2-rank of the class group. Acta Arith. **182**, 117–132 (2018)
23. S. Kuroda, On the class number of imaginary quadratic number fields. Proc. Jpn. Acad. **40**, 365–367 (1964)
24. A. Levin, Ideal class groups, Hilbert's irreducibility theorem, and integral points of bounded degree on curves. J. Théor. Nombres Bordeaux **19**(2), 485–499 (2007)
25. J.-F. Mestre, Courbes elliptiques et groupes de classes d'idéaux de certains corps quadratiques, in *Seminar on Number Theory, 1979–1980* (French) (Univ. Bordeaux I, Talence, 1980), pp. Exp. No. 15, 18
26. J.-F. Mestre, Courbes elliptiques et groupes de classes d'idéaux de certains corps quadratiques. J. Reine Angew. Math. **343**, 23–35 (1983)
27. J.-F. Mestre, Groupes de classes d'idéaux non cycliques de corps de nombres, in *Seminar on number theory, Paris 1981–82 (Paris, 1981/1982)*, Progr. Math., vol. 38 (Birkhäuser Boston, Boston, MA, 1983), pp. 189–200
28. J.-F. Mestre, Corps quadratiques dont le 5-rang du groupe des classes est ≥ 3. C. R. Acad. Sci. Paris Sér. I Math. **315**(4), 371–374 (1992)
29. T. Nagell, Über die Klassenzahl imaginär-quadratischer Zahlkörper. Abh. Math. Sem. Univ. Hambg. **1**, 140–150 (1922)
30. T. Nagell, Contributions to the theory of a category of Diophantine equations of the second degree with two unknowns. Nova Acta Sci. Soc. Upsal. Ser. (4) **16**, 1–38 (1955)
31. T. Nagell, Collected papers of Trygve Nagell, vol. 1, in *Queen's Papers in Pure and Applied Mathematics*, vol. 121 (Queen's University, Kingston, ON, 2002)
32. S. Nakano, On ideal class groups of algebraic number fields. Proc. Jpn. Acad. Ser. A Math. Sci. **60**(2), 74–77 (1984)
33. S. Nakano, On ideal class groups of algebraic number fields. J. Reine Angew. Math. **358**, 61–75 (1985)
34. S. Nakano, Construction of pure cubic fields with large 2-class groups. Osaka J. Math. **25**(1), 161–170 (1988)
35. M.R. Murty, The ABC conjecture and exponents of class groups of quadratic fields. Contemp. Math. **210**, 85–95 (1998)
36. M.R. Murty, Exponents of class groups of quadratic fields, in *Topics in Number Theory*, vol. 467 (University Park, PA, 1997), pp. 229–239. Math. Appl., Kluwer Acad. Publ. Dordrecht, 1999
37. K. Soundararajan, Divisibility of class numbers of imaginary quadratic fields. J. Lond. Math. Soc. (2) **61** (3), 681–690 (2000)
38. K. Uchida, Class numbers of cubic cyclic fields. J. Math. Soc. Jpn. **26**(3), 447–453 (1974)
39. P.J. Weinberger, Real quadratic fields with class numbers divisible by n. J. Number Theory **5**(3), 237–241 (1973)
40. Y. Yamamoto, On unramified Galois extensions of quadratic number fields. Osaka J. Math. **7**, 57–76 (1970)
41. G. Yu, A note on the divisibility of class numbers of real quadratic fields. J. Number Theory **97**(1), 35–44 (2002)

On Simultaneous Divisibility of the Class Numbers of Imaginary Quadratic Fields

Toru Komatsu

1 Introduction

In this article, we explain some results in the papers [7, 8] on simultaneous divisibility of the class numbers of quadratic fields and present an evolved problem of the inverse Galois problem.

Let k be an algebraic number field with $[k : \mathbb{Q}] < \infty$. Let $Cl(k)$ denote the ideal class group of k, and $h(k)$ the class number of k.

Theorem 1.1 (Komatsu [7] 2002, Acta Arith.) *Let $m \neq 0$ be a rational integer. Then there exist infinitely many real (imaginary) quadratic fields $\mathbb{Q}(\sqrt{D})$ such that $3 \mid h(\mathbb{Q}(\sqrt{D}))$ and $3 \mid h(\mathbb{Q}(\sqrt{mD}))$.*

Theorem 1.2 (Komatsu [8] 2017, IJNT) *Let n and m be rational integers greater than 1. Then, there exist infinitely many imaginary quadratic fields $\mathbb{Q}(\sqrt{D})$ such that $n \mid h(\mathbb{Q}(\sqrt{D}))$ and $n \mid h(\mathbb{Q}(\sqrt{mD}))$.*

2 Old Motivation for the Results

Let $d > 1$ be a squarefree rational integer. Let r denote the 3-rank of $Cl(\mathbb{Q}(\sqrt{d}))$ of the real quadratic field $\mathbb{Q}(\sqrt{d})$, and s that of the imaginary quadratic field $\mathbb{Q}(\sqrt{-3d})$.

Theorem 2.1 (Scholz, reflection theorem) *The inequality $r \leq s \leq r + 1$ holds. For a rational integer $d > 1$, if $3 \mid h(\mathbb{Q}(\sqrt{d}))$, then $3 \mid h(\mathbb{Q}(\sqrt{-3d}))$.*

The conference ICCGNFRT at Harish-Chandra Research Institute on September 2017.

T. Komatsu (✉)
Tokyo University of Science,Tokyo, Japan
e-mail: komatsu_toru@ma.noda.tus.ac.jp

K. Chakraborty et al. (eds.), *Class Groups of Number Fields and Related Topics*,
https://doi.org/10.1007/978-981-15-1514-9_2

Remark 2.2 By some experiments with calculators, I wondered if the 3-divisibilities of $h(\mathbb{Q}(\sqrt{d}))$ and $h(\mathbb{Q}(\sqrt{-d}))$ are independent of one another. The problem is whether there are infinitely many quadratic fields $\mathbb{Q}(\sqrt{D})$ such that $3 \mid h(\mathbb{Q}(\sqrt{D}))$ and $3 \mid h(\mathbb{Q}(\sqrt{-D}))$ or not. It was solved affirmatively in a paper at 2001. Its general cases are done in [7] at 2002.

3 Comparison of Methods

Let $H(k)$ denote the Hilbert class field of k, that is, the maximal unramified abelian extension of k. Class field theory yields an isomorphism $Cl(k) \simeq \mathrm{Gal}(H(k)/k)$ where $\mathrm{Gal}(H(k)/k)$ is the Galois group of the extension $H(k)/k$.

Remark 3.1 In the paper [7] (2002), we construct unramified cyclic cubic extensions of k due to Honda's method [4] (1968) and also to Kishi-Miyake (2000). It is not necessary for the method to consider influence of units. In the paper [8] (2017), we construct ideals of k with order n in $Cl(k)$ due to Yamamoto's method [12] (1970), which needs consideration for the influence of units.

4 Construction of Fields and Extensions

Let us recall the result in the paper [7]. Let $m \neq 1$ be a squarefree rational integer. Let l be a prime number which splits in the extension $\mathbb{Q}(\sqrt{m})/\mathbb{Q}$ and is inert in the extension $\mathbb{Q}(\sqrt[3]{2})/\mathbb{Q}$. We take a rational integer ν such that

$$\nu \equiv \begin{cases} \pm(4m-3) \pmod{27} & \text{if } m \equiv 1 \pmod{3}, \\ \pm(4m+12) \pmod{27} & \text{if } m \equiv 2 \pmod{3}, \\ \pm 4m \pmod{27} & \text{if } m \equiv 3 \pmod{9}, \\ \pm 1 \pmod{3} & \text{otherwise}, \end{cases}$$

and $m\nu^2 \equiv 1 \pmod{l}$. Now put $r = m\nu^2$. Let T be the set of all of the rational integers t such that

$$t \equiv \begin{cases} 4 \text{ or } 7 \pmod{9} & \text{if } m \equiv 1 \pmod{3}, \\ 3 \pmod{9} & \text{if } m \equiv 2 \pmod{3}, \\ -3 \pmod{27} & \text{if } m \equiv 3 \pmod{9}, \\ \pm(r/3)^2 \pmod{9} & \text{otherwise}, \end{cases}$$

$t \equiv -1 \pmod{l}$ and $t \not\equiv r \pmod{p}$ for every prime divisor $p \neq 3$ of $r(r-1)$. We define

$$D_r(X) := (3X^2 + r)(2X^3 - 3(r+1)X^2 + 6rX - r(r+1))/27.$$

Theorem 4.1 (Komatsu [7]) *For each $t \in T$, we have that $3 \mid h(\mathbb{Q}(\sqrt{D_r(t)}))$ and $3 \mid h(\mathbb{Q}(\sqrt{mD_r(t)}))$. When $m > 0$, if $t \geq 3r/2$ (resp. $t < 3r/2$), then $\mathbb{Q}(\sqrt{D_r(t)})$ and $\mathbb{Q}(\sqrt{mD_r(t)})$ are both real (resp. both imaginary).*

For $t \in T$, put

$$u := t^3 + 3tr, \quad w := 3t^2 + r, \quad a := u - w, \quad b := u - rw, \quad c := t^2 - r.$$

We define

$$f_1(Z) := Z^3 - 3cZ - 2a, \quad f_2(Z) := Z^3 - 3cZ - 2b.$$

For $j = 1$ and 2, let K_j denote the minimal splitting field of $f_j(Z)$ over $\mathbb{Q}$, and $d(f_j)$ the discriminant of the polynomial $f_j(Z)$. Put $k_j := \mathbb{Q}(\sqrt{d(f_j)})$.

Proposition 4.2 *For every $j = 1$ and 2, the extension K_j/k_j is cyclic cubic and unramified. We have that $k_1 = \mathbb{Q}(\sqrt{D_r(t)})$ and $k_2 = \mathbb{Q}(\sqrt{mD_r(t)})$.*

Proof By the definition one has that $r \equiv 1 \pmod{l}, t \equiv -1 \pmod{l}, a \equiv b \equiv -2^3 \pmod{l}$ and $c \equiv 0 \pmod{l}$. Note that $2 \notin \mathbb{F}_l^3$. Thus $f_j \equiv Z^3 + 2^4 \pmod{l}$ are irreducible over $\mathbb{F}_l$, and so are over $\mathbb{Q}$. Thus K_j/k_j are cyclic cubic.

By using Llorente–Nart's criterion for the decompositions of primes in cubic fields $\mathbb{Q}(\theta)$ where $f_j(\theta) = 0$, we see that $\mathbb{Q}(\theta)/\mathbb{Q}$ are not totally ramified at any finite primes. Hence K_j/k_j are unramified. □

Remark 4.3 The construction yields infinite families, not only of pairs of imaginary and imaginary, but also of those of real and real.

Remark 4.4 Without considering any influence of units, we focus on only the sign of the discriminant. In general, for the n-divisibilities of the class numbers of quadratic fields, we may try to construct $\mathcal{D}_n$-extensions of $\mathbb{Q}$ as the minimal splitting fields of polynomial with degree n. It is difficult to make such polynomials of large degree n with parameters yielding infinite family.

5 Construction of Fields and Ideals

Let us explain the result in the paper [8]. Let $n > 1$ be a rational integer with the prime decomposition $n = p_1^{e_1} p_2^{e_2} \dots p_s^{e_s}$. Let $m > 1$ be a squarefree rational integer.
Step 1. We take distinct prime numbers l_i such that $l_i \equiv 1 \pmod{12p_i}$ and $m \in \mathbb{F}_{l_i}^{\times 2}$ for $i = 1, 2, \dots, s$.
Step 2. For each i, we take a rational integer g_i with $g_i, g_i + 1 \notin \mathbb{F}_{l_i}^{p_i}$.
Step 3. We take a positive even number a such that $a^2 m \equiv g_i^2/(g_i + 1)^2 \pmod{l_i}$ for all i.
Step 4. We take a rational integer t satisfying all of the following conditions:

$$\begin{cases} t \equiv g_i/(g_i+1) \pmod{l_i} \text{ for all } i, \\ \gcd(t, am) = 1, \\ \gcd(t-1, b) = 1, \\ t > a^2mn/2, \end{cases}$$

where b is the maximal divisor of $a^2m - 1$ relatively prime to $l_1 l_2 \cdots l_s$.
Step 5. We put $M := \mathbb{Q}(\sqrt{m})$, and

$$\beta := t - a\sqrt{m}, \quad \gamma_1 := 1 + \frac{1}{a\sqrt{m}}, \quad \gamma_2 := 1 + a\sqrt{m},$$
$$x_1 := \mathrm{Tr}(\beta^n \gamma_1), \quad x_2 := \mathrm{Tr}(\beta^n \gamma_2), \quad z_1 := z_2 := N(\beta),$$

where $\mathrm{Tr} = \mathrm{Tr}_{M/\mathbb{Q}}$ and $N = N_{M/\mathbb{Q}}$. For each $j = 1$ and 2, let F_j denote the quadratic field $\mathbb{Q}(\sqrt{x_j^2 - 4z_j^n})$ with discriminant D_j.

Theorem 5.1 (Komatsu [8]) *For each $j = 1$ and 2, F_j is an imaginary quadratic field with an ideal of order n. The ratio $D_2/(mD_1)$ is square.*

Remark 5.2 For each $j = 1$ and 2, there exists a rational integer y_j such that $x_j^2 - 4z_j^n = y_j^2 D_j$. We put $\alpha_j := (x_j + y_j\sqrt{D_j})/2$. Let $\mathfrak{a}_j$ be an ideal of F_j satisfying $\mathfrak{a}_j^n = (\alpha_j)$. This implies that $\mathfrak{a}_j$ is the ideal generated by z_j and α_j. Then the order of $\mathfrak{a}_j$ in $Cl(F_j)$ is equal to n.

For the existence of l_i, we have the following lemma.

Lemma 5.3 *Let p be a prime number. If l is a prime number with $l \equiv 1 \pmod{12mp}$, then $l \equiv 1 \pmod{12p}$ and $m \in \mathbb{F}_l^{\times 2}$.*

Proof Let m_0 be the maximal odd divisor of m. Then one has $\left(\frac{m}{l}\right) = \left(\frac{m_0}{l}\right) = \left(\frac{l}{m_0}\right) = \left(\frac{1}{m_0}\right) = 1.$ □

For the existence of g_i, we have the following lemma.

Lemma 5.4 *Let p and l be prime numbers with $l \equiv 1 \pmod{2p}$. Then there exists a rational integer g such that both g and $g+1$ are pth power non-residue modulo l.*

Proof Note that 1 and $l - 1$ are pth power residues modulo l. Let g be a pth power non-residue modulo l with $2 \le g \le l - 2$. Assume that $g + 1$ is a pth power residue modulo l. Then g^{-1} is a suitable one, that is, g^{-1} and $g^{-1} + 1 = (g+1)/g$ are pth power non-residue modulo l. □

Remark 5.5 We can show the existence of a and t due to Chinese remainder theorem (CRT). Since l_i are odd, distinct and $\left(\frac{m}{l_i}\right) = 1$, the integer a proves to exist by CRT. Since $l_1, \ldots, l_s, am, b$ are relatively prime to each other, the integer t proves to exist by CRT.

Remark 5.6 Using Yamamoto's method [12], one can see that $\mathfrak{a}_j$ is of order n in $Cl(F_j)$. In fact, by $\mathfrak{a}_j^n = (\alpha_j)$, the order of $\mathfrak{a}_j$ is a divisor of n. Since α_j is a p_ith power non-residue modulo $\mathfrak{l}_i$ above l_i and units are power residues, the order does not decrease.

Remark 5.7 With l_i, g_i, and a fixed, the family of fields constructed in the run of t is infinite. Indeed, for every large number C, the family contains a quadratic field ramified at a prime number greater than C.

6 New Motivation, Application to a Problem

Let k be a number field with $[k : \mathbb{Q}] < \infty$, and K a Galois extension of k with $[K : k] < \infty$. Let G denote the Galois group of K/k, and $\mathcal{H}$ the family of all of the subgroups of G, that is,

$$\mathcal{H} := \{H : \text{ subgroups of } G\} := \{H_1, H_2, \ldots, H_s\},$$

where s is the number of the subgroups of G. Let K_j denote the fixed field by H_j in K/k. Let $\mathfrak{q}$ be an integral ideal of K, and $\mathfrak{q}_j$ the ideal of K_j below $\mathfrak{q}$, that is, $\mathfrak{q} \cap K_j$. Let n_j denote the order of $\mathfrak{q}_j$ in $Cl(K_j)$, respectively.

Definition 6.1 *We say that $(n_1, n_2, \ldots, n_s)$ is the tuple of the orders in the extension K/k of $\mathfrak{q}$, and shorten it to toe of $\mathfrak{q}$.*

Definition 6.2 *(Inverse Galois problem with toe condition) Let k be a number field with $[k : \mathbb{Q}] < \infty$. Let G be a finite group and $\mathcal{H}$ the family of all of the subgroups of G, $\mathcal{H} := \{H : \text{ subgroups of } G\} := \{H_1, H_2, \ldots, H_s\}$. For given positive integers $n_1, n_2, \ldots, n_s$, does there exist a Galois G-extension K of k with an ideal of toe $(n_1, n_2, \ldots, n_s)$?*

Remark 6.1 It seems to need some conditions on n_j's according to the relations between H_j's.

Let $k = \mathbb{Q}$ and $G = \{e, \sigma, \tau, \sigma\tau\} \simeq V_4 \simeq C_2 \times C_2$ with

$$\mathcal{H} = \{H_1 = \{e\}, H_2 = \langle\sigma\rangle, H_3 = \langle\tau\rangle, H_4 = \langle\sigma\tau\rangle, H_5 = G\}.$$

Corollary 6.3 *Let b, c, d be positive, odd numbers and put $a := lcm(b, c, d)$. Then there exist infinitely many Galois V_4-extensions K of $\mathbb{Q}$ with ideals $\mathfrak{a}$ of toe $(a, b, c, d, 1)$.*

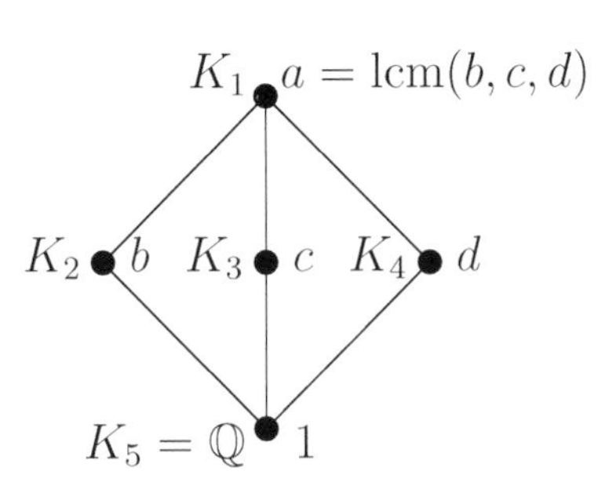

Proof By Yamamoto's result, there exists a real quadratic field $\mathbb{Q}(\sqrt{m})$ with an ideal $\mathfrak{b}$ of order b. Due to Theorem 5.1 at Sect. 5, there exist imaginary quadratic fields $\mathbb{Q}(\sqrt{D})$ and $\mathbb{Q}(\sqrt{mD})$ with ideals $\mathfrak{c}$ and $\mathfrak{d}$ of order cd, respectively. Put $K = \mathbb{Q}(\sqrt{m}, \sqrt{D})$ and $G = \mathrm{Gal}(K/\mathbb{Q}) = \langle \sigma, \tau \rangle$ where

$$\sigma : \sqrt{m} \mapsto \sqrt{m}, \sqrt{D} \mapsto -\sqrt{D},$$
$$\tau : \sqrt{m} \mapsto -\sqrt{m}, \sqrt{D} \mapsto \sqrt{D}.$$

Then it follows from the Galois correspondence that $K_1 = K$, $K_2 = \mathbb{Q}(\sqrt{m})$, $K_3 = \mathbb{Q}(\sqrt{D})$, $K_4 = \mathbb{Q}(\sqrt{mD})$, and $K_5 = \mathbb{Q}$. Put $\mathfrak{a} = \mathfrak{b}\mathfrak{c}^d\mathfrak{d}^c$ as ideals of K. Let $\mathfrak{q}$ be a prime ideal of K which is equivalent to $\mathfrak{a}$ in $Cl(K)$, and which splits completely in $K/\mathbb{Q}$. Then the toe of $\mathfrak{q}$ is $(a, b, c, d, 1)$. Indeed, the classes $[\mathfrak{q}_j]$ of $\mathfrak{q}_j$ in $Cl(K_j)$ are as follows:

$$[\mathfrak{q}_1] = [\mathfrak{q}] = [\mathfrak{a}], \quad [\mathfrak{q}_2] = [N_{K/K_2}(\mathfrak{q})] = [\mathfrak{b}^2],$$
$$[\mathfrak{q}_3] = [N_{K/K_3}(\mathfrak{q})] = [\mathfrak{c}^{2d}], \quad [\mathfrak{q}_4] = [N_{K/K_4}(\mathfrak{q})] = [\mathfrak{d}^{2c}], \quad [\mathfrak{q}_5] = [(q)].$$

This completes the proof. □

7 Real Quadratic Cases

Let us recall Yamamoto's method in [12] not only for the imaginary but also for the real. Let F be a quadratic field with discriminant $D \neq -3, -4$. Let x, y, z be rational integers such that $x^2 - y^2D = 4z^n$ and $\gcd(x, z) = 1$. Put $\alpha_\pm := (x \pm y\sqrt{D})/2$. Then there exists an ideal $\mathfrak{a}$ of F such that $\mathfrak{a}^n = (\alpha_+)$. Put

$$\varepsilon := \begin{cases} \text{the fundamental unit of } F & \text{if } D > 0, \\ 1 & \text{if } D < -4. \end{cases}$$

Let p be a prime factor of n, and l a prime number with $l \equiv 1 \pmod{2p}$. Assume that $x \notin \mathbb{F}_l^p$ and $l \mid z$. Then there exists a prime ideal $\mathfrak{l}$ of F above l dividing α_-.

Lemma 7.1 (Yamamoto [12]) *If ε is a pth power residue modulo $\mathfrak{l}$, then (α_+) is the pth power of no principal ideal in F.*

Let n be a rational integer greater than 1 with the prime decomposition $n = p_1^{e_1} p_2^{e_2} \dots p_s^{e_s}$. Let l_i, l_i' be distinct prime numbers with $l_i \equiv l_i' \equiv 1 \pmod{2p_i}$. Let x, z, x', z' be rational integers such that

$$\begin{cases} x^2 - 4z^n = x'^2 - 4z'^n, \\ \gcd(x, z) = \gcd(x', z') = 1, \\ x \notin \mathbb{F}_{l_i}^{p_i}, \quad x' \notin \mathbb{F}_{l_i'}^{p_i}, \quad \dfrac{x + x'}{2} \in \mathbb{F}_{l_i}^{p_i}, \\ l_i \mid z, \quad l_i' \mid z'. \end{cases}$$

Let F denote the quadratic field $\mathbb{Q}(\sqrt{x^2 - 4z^n})$ with discriminant D.

Theorem 7.2 (Yamamoto [12], Prop. 2) *The ideal class group $Cl(F)$ has a subgroup $\mathcal{N}$ isomorphic to $C_n \times C_n$ if $D < -4$, and C_n if $D > 0$.*

Proof There exists a rational integer y such that $x^2 - 4z^n = x'^2 - 4z'^n = y^2 D$. Then there exist ideals $\mathfrak{a}, \mathfrak{a}'$ of F such that $\mathfrak{a}^n = ((x + y\sqrt{D})/2)$ and $\mathfrak{a}'^n = ((x' + y\sqrt{D})/2)$. When $D < -4$, the ideals $\mathfrak{a}$ and $\mathfrak{a}'$ generate $\mathcal{N} \simeq C_n \times C_n$. When $D > 0$, the ideals $\mathfrak{a}$ and $\mathfrak{a}'$ may have orders less than n because of units in F; however, $\mathfrak{a}$ and $\mathfrak{a}'$ generate an ideal of order n in $Cl(F)$. □

Remark 7.3 Diophantine equation $X^2 - Y^2 D = 4Z^n$ does not become complicated as n increases. We need to consider the influence of units. The number of using integers $x, \dots$ for real quadratic cases is more than that for imaginary quadratic cases.

Acknowledgements The author would like to express his deepest gratitude to the organizers Prof. Kalyan Chakraborty, Dr. Azizul Hoque, and Dr. Prem Prakash Pandey for giving an opportunity to talk and inviting to the conference ICCGNFRT held at Harish-Chandra Research Institute on September 2017.

References

1. N.C. Ankeny, S. Chowla, On the divisibility of the class number of quadratic fields. Pac. J. Math. **5**, 321–324 (1955)
2. P. Hartung, Proof of the existence of infinitely many imaginary quadratic fields whose class number is not divisible by 3. J. Number Theory **6**, 276–278 (1974)
3. P. Hartung, Explicit construction of a class of infinitely many imaginary quadratic fields whose class number is divisible by 3. J. Number Theory **6**, 279–281 (1974)
4. T. Honda, On real quadratic fields whose class numbers are multiples of 3. J. Reine Angew. Math. **233**, 101–102 (1968)
5. P. Humbert, Sur les nombres de classes de certains corps quadratiques. Comment. Math. Helv. **12**, 233–245 (1940)

6. A. Ito, Existence of an infinite family of pairs of quadratic fields $\mathbb{Q}(\sqrt{m_1 D})$ and $\mathbb{Q}(\sqrt{m_2 D})$ whose class numbers are both divisible by 3 or both indivisible by 3. Funct. Approx. Comment. Math. **49**(1), 111–135 (2013)
7. T. Komatsu, An infinite family of pairs of quadratic fields $\mathbb{Q}(\sqrt{D})$ and $\mathbb{Q}(\sqrt{mD})$ whose class numbers are both divisible by 3. Acta Arith. **104**(2), 129–136 (2002)
8. T. Komatsu, An infinite family of pairs of imaginary quadratic fields with ideal classes of a given order. Int. J. Number Theory **13**, 253–260 (2017)
9. S. Kuroda, On the class number of imaginary quadratic number fields. Proc. Jpn. Acad. **40**, 365–367 (1964)
10. T. Nagel, Über die Klassenzahl imaginär-quadratischer Zahlkörper. Abh. Math. Sem. Univ. Hambg. **1**, 140–150 (1922)
11. P.J. Weinberger, Real quadratic fields with class numbers divisible by n. J. Number Theory **5**, 237–241 (1973)
12. Y. Yamamoto, On unramified Galois extensions of quadratic number fields. Osaka J. Math. **7**, 57–76 (1970)

Thue Diophantine Equations

A Survey

Michel Waldschmidt

Mathematics Subject Classification (2000) MSC 11D59

1 Thue Equations

1.1 Introduction

A *Thue equation* is a Diophantine equation of the form $F(x, y) = m$, where $F \in \mathbb{Z}[X, Y]$ is a given homogeneous polynomial in two variables (i.e., a binary form) of degree d with integer coefficients, m is a given nonzero integer while the unknowns x, y take their values in $\mathbb{Z}$. Is the set of such (x, y) finite or infinite? If it is finite, can we get an upper bound for the number of its elements? (Such an upper bound is a qualitative statement). Can we get an upper bound for the height of its elements? (Such an upper bound is a quantitative statement).

A *Thue–Mahler equation* is an exponential Diophantine equation of the form $F(x, y) = p_1^{z_1} \cdots p_s^{z_s}$ where F is a given binary form, $p_1, \ldots, p_s$ are given prime numbers, the unknowns are $x, y, z_1, \ldots, z_s$ where x, y take their values in $\mathbb{Z}$ and $z_1, \ldots, z_s$ in $\mathbb{Z}_{\geq 0}$.

M. Waldschmidt (✉)
Sorbonne Université, Université Pierre et Marie Curie, CNRS,
Institut de Mathématiques de Jussieu–Paris Rive Gauche, IMJ–PRG,
75005 Paris, France
e-mail: michel.waldschmidt@imj-prg.fr
URL: http://www.imj-prg.fr/~michel.waldschmidt

K. Chakraborty et al. (eds.), *Class Groups of Number Fields and Related Topics*,
https://doi.org/10.1007/978-981-15-1514-9_3

We denote by $f \in \mathbb{Z}[T]$ the polynomial defined by $f(T) = F(T, 1)$:

$$f(T) = a_0 T^d + a_1 T^{d-1} + \cdots + a_{d-1} T + a_d,$$
$$F(X, Y) = a_0 X^d + a_1 X^{d-1} Y + \cdots + a_{d-1} X Y^{d-1} + a_d Y^d.$$

Notice that $a_0 = 0$ is equivalent to saying that $F(X, 0)$ is the zero polynomial. We assume $a_0 > 0$ so that f has degree d.

For $m = 0$, the set of $(x, y) \neq (0, 0)$ in $\mathbb{Z}^2$ such that $F(x, y) = 0$ is empty if f has no rational root, while, if f has rational roots, then this set is the set of (x, y) with $y \neq 0$ such that x/y is a root of f.

From now on we assume $m \neq 0$.

When $d = 1$, we have $F(X, Y) = a_0 X + a_1 Y$; the solution of a linear equation $a_0 x + a_1 y = m$ is given by Bézout's Theorem. The computation of the gcd of a_0 and a_1 is done efficiently via the Euclidean algorithm, which is nothing else than the continued fraction expansion algorithm applied to a_1/a_0.

Assume $d = 2$. The quadratic equation $a_0 x^2 + a_1 xy + a_2 y^2 = m$ may have no solution or finitely many solutions: one among many examples is for $(a_0, a_1, a_2) = (1, 0, 1)$ with the equation $x^2 + y^2 = m$. It may have infinitely many solutions; this is the case for $(a_0, a_1, a_2) = (1, 0, -D)$ and $m = 1$, where D is a positive integer which is not a square, with the Brahmagupta–Fermat–Pell equation $x^2 - Dy^2 = 1$. The general solution of the quadratic equation (not necessarily a Thue equation) is due to Lagrange [11, 28].

Assume now $d > 2$. If f is a reducible polynomial in $\mathbb{Z}[X]$, then solving the equation $F(x, y) = m$, where the unknowns x, y take their values in the set of rational integers, amounts to solving finitely many equations $F_i(x, y) = m_i$ with m_i a divisor of m and $F_i(X, Y)$ an irreducible factor of $F(X, Y)$ in $\mathbb{Z}[X, Y]$. For this reason, we assume now that f is irreducible in $\mathbb{Z}[X]$.

1.2 *Positive Definite Binary Forms*

Assume first that the polynomial f has no real root (hence its degree d is even). Then for each $m \in \mathbb{Z}$, $m \neq 0$, the set of (x, y) in $\mathbb{Z}^2$ such that $F(x, y) = m$ is finite. To study the Diophantine equation $F(x, y) = m$ means to study the representation of integers by the definite form F. Let us quote the following elementary Lemma 2.1 from [12].

Lemma 1 *Let $f \in \mathbb{Z}[T]$ be a nonzero polynomial of degree d which has no real root. Let $g(T) = T^d f(1/T)$. Assume that the leading coefficient of $f(T)$ is positive so that the real number, defined by*

$$\gamma = \min\left\{ \inf_{-1 \le t \le 1} f(t), \quad \inf_{-1 \le t \le 1} g(t) \right\}$$

is > 0. *Let* $F(X, Y)$ *be the binary form* $Y^d f(X/Y)$ *associated with* f. *Then for each* $(x, y) \in \mathbb{Z}^2$, *we have*

$$F(x, y) \geq \gamma \max\{|x|^d, |y|^d\}.$$

Moreover, for any real number c *with* $c > \gamma$, *there exist infinitely many couples* (x, y) *in* $\mathbb{Z}^2$ *satisfying*

$$F(x, y) < c \max\{|x|^d, |y|^d\}.$$

A class of definite forms F (namely, forms which are associated with a polynomial f without real root) is given by the norm over $\mathbb{Q}$ of a CM field. Recall that a subfield K of $\mathbb{C}$ is a CM field if is number field which satisfies the following equivalent conditions:
(i) K is totally imaginary and is a quadratic extension of a totally real field.
(ii) There exists $\gamma \in K$ such that $K = \mathbb{Q}(\gamma)$ and γ^2 is totally real with all conjugates negative.
(iii) K is not real and the complex conjugation $z \to \overline{z}$ commutes with every embedding of K into $\mathbb{C}$: for $\sigma : K \to \mathbb{C}$ and $\alpha \in K$,

$$\overline{\sigma(\alpha)} = \sigma(\overline{\alpha}).$$

Theorem 1 (Győry [13]) *Let* K *be a CM field of degree* d *over* $\mathbb{Q}$. *Let* $\alpha \in K$ *be such that* $K = \mathbb{Q}(\alpha)$; *let* f *be the irreducible polynomial of* α *over* $\mathbb{Q}$ *and let* $F(X, Y) = Y^d f(X/Y)$ *the associated homogeneous binary form. Set* $a_0 = F(1, 0)$, $a_d = F(0, 1)$. *For* $(x, y) \in \mathbb{Z}^2$ *we have*

$$x^d \leq 2^d a_d^{d-1} F(x, y) \quad \textit{and} \quad y^d \leq 2^d a_0^{d-1} F(x, y).$$

Recall that the leading coefficient a_0 of the irreducible polynomial of an algebraic number is positive. The assumption implies that α is totally imaginary, hence $a_d > 0$ and $F(x, y) > 0$ for $(x, y) \neq (0, 0)$.

Proof Let $\alpha_1, \ldots, \alpha_d$ be the roots of f in $\mathbb{C}$ so that

$$F(X, Y) = a_0(X - \alpha_1 Y) \cdots (X - \alpha_d Y).$$

For $1 \leq j \leq d$, the number α_j is not real (since K is totally imaginary) and we have

$$|x - \alpha_j y| \geq |\Im\mathrm{m}(\alpha_j) y|.$$

Since K is a CM field, $\overline{\alpha}$ is in K and $2i\Im\mathrm{m}(\alpha) = \alpha - \overline{\alpha}$ is a nonzero element in K; its conjugates in $\mathbb{C}$ are $\alpha_j - \overline{\alpha_j}$. Moreover, $a_0(\alpha - \overline{\alpha})$ being a nonzero algebraic integer, its norm is a nonzero rational integer, of absolute value ≥ 1. Therefore

$$2^d a_0^{d-1} F(x, y) = y^d \mathrm{N}_{K/\mathbb{Q}}(2ia_0 \Im\mathrm{m}(\alpha)) \geq y^d.$$

The same argument gives the upper bound for x^d. □

In the special case where α is a unit in K, we have $a_0 = a_d = 1$ and the conclusion can be written as

$$\max\{|x|, |y|\} \leq 2F(x, y)^{1/d}. \quad (1)$$

Examples of binary forms satisfying the assumptions of Theorem 1 with $a_0 = a_d = 1$ are given by the cyclotomic binary forms, which we define as follows.

For $n \geq 1$, denote by $\phi_n(T)$ the cyclotomic polynomial of index n and degree $\varphi(n)$ (Euler's totient function). The *cyclotomic binary form* $\Phi_n(X, Y)$ is defined by $\Phi_n(X, Y) = Y^{\varphi(n)}\phi_n(X/Y)$. In particular, we have $\Phi_n(x, y) > 0$ for $n \geq 3$ and $(x, y) \neq (0, 0)$.

An example showing that the estimate (1) is optimal is given by the form $F(X, Y) = \Phi_n(X - Y, Y)$ of degree $d = \varphi(n)$, where $n \geq 3$ is not of the form p^r nor $2p^r$ with p prime. This condition on n implies $\phi_n(1) = \phi_n(-1) = 1$, hence for $y \in \mathbb{Z}$ we have

$$F(2y, y) = y^d F(2, 1) = y^d \phi_n(1) = y^d.$$

The irreducible polynomial of the unit $\alpha = 1 + \zeta_n$ is $\phi_n(t - 1)$ and the field $\mathbb{Q}(\alpha)$ is the CM field $\mathbb{Q}(\zeta_n)$.

In the special case of cyclotomic binary forms, Theorem 1 gives

$$\max\{|x|, |y|\} \leq 2|m|^{1/\varphi(n)}$$

for the integral solutions (n, x, y) of $\Phi_n(x, y) = m$. An upper bound for n can be deduced only if $\max\{|x|, |y|\} \geq 3$.

In [12], the refined estimate

$$\max\{|x|, |y|\} \leq \frac{2}{\sqrt{3}}|m|^{1/\varphi(n)}$$

has been proved for the integral solutions (n, x, y) of $\Phi_n(x, y) = m$ satisfying $n \geq 3$ and $\max\{|x|, |y|\} \geq 2$. Therefore

$$\varphi(n) \leq \frac{2}{\sqrt{3}} \log m.$$

See [29], [A296095, A299214, A293654, A301429, and A301430].

1.3 Thue Equation and Diophantine Approximation

We now come back to the general case where $f \in \mathbb{Z}[T]$ is irreducible over $\mathbb{Q}$ of degree $d \geq 3$ and may have real zeroes. Write

$$f(T) = a_0(T - \alpha_1)(T - \alpha_2) \cdots (T - \alpha_d).$$

Recall our assumption $a_0 > 0$. Assume $m \neq 0$ is fixed while x, y are rational integers with $F(x, y) = m$. Let us show that, as soon as $\max\{|x|, |y|\}$ is sufficiently large, x/y is close to one of the roots α_i of f and is not close to any other root (since f is irreducible, the roots α_i are distinct). For $i = 1, \ldots, d$, define $\beta_i = x - \alpha_i y$. Label the roots of f so that

$$|\beta_1| = \min_{1 \leq i \leq d} |\beta_i|.$$

From $a_0\beta_1 \ldots \beta_d = m$ we deduce

$$|\beta_1|^d \leq \frac{|m|}{a_0},$$

which means

$$\left|\alpha_1 - \frac{x}{y}\right| \leq \frac{|m|^{1/d}}{a_0^{1/d}|y|}. \tag{2}$$

We may notice that if α_1 is not real, then we get immediately a sharp upper bound for $|y|$:

$$|y| \leq \frac{|m|^{1/d}}{a_0^{1/d}|\Im\mathrm{m}(\alpha_1)|}.$$

We now sharpen the upper bound (2). If

$$|y| \leq \frac{2|m|^{1/d}}{a_0^{1/d} \min\limits_{2 \leq i \leq d} |\alpha_i - \alpha_1|},$$

then the relation $x = \alpha_1 y + \beta_1$ implies the upper bound

$$|x| \leq \left(\frac{2|\alpha_1|}{\min\limits_{2 \leq i \leq d} |\alpha_i - \alpha_1|} + 1\right) \frac{|m|^{1/d}}{a_0^{1/d}},$$

which shows that $|x|$ and $|y|$ are bounded. Assume now

$$|y| > \frac{2|m|^{1/d}}{a_0^{1/d} \min\limits_{2 \leq i \leq d} |\alpha_i - \alpha_1|}.$$

For $i = 2, \ldots, d$, we have $\beta_i = (\alpha_1 - \alpha_i)y + \beta_1$, hence

$$|\beta_i| = |x - \alpha_i y| \geq |(\alpha_i - \alpha_1)y| - \frac{|m|^{1/d}}{a_0^{1/d}} \geq \frac{1}{2}|(\alpha_i - \alpha_1)y|,$$

which implies

$$|m| = |a_0\beta_1 \ldots \beta_d| \geq |y|^{d-1}|\beta_1|\frac{1}{2^{d-1}}a_0\prod_{i=2}^{d}|\alpha_i - \alpha_1|$$

and therefore we deduce the following improvement of the upper bound (2):

$$\left|\alpha_1 - \frac{x}{y}\right| \leq \frac{\kappa_1(f)|m|}{|y|^d}$$

with

$$\kappa_1(f) = \frac{2^{d-1}}{a_0\prod_{i=2}^{d}|\alpha_i - \alpha_1|}.$$

Liouville's estimate is the lower bound

$$\left|\alpha_1 - \frac{x}{y}\right| \geq \frac{\kappa_2(f)}{|y|^d}$$

with some explicit constant $\kappa_2(f)$; this does not give any information on the Thue Diophantine equation, but any nontrivial improvement of Liouville's estimate gives a nontrivial information on the equation $F(x, y) = m$ (see [18, 35]). The work by Thue in 1914 culminated with the proof by Roth in 1955 of the following result.

Theorem 2 (Thue, Siegel, Roth) *If α is an algebraic number of degree $d \geq 2$, for any $\varepsilon > 0$ there exists a positive constant $\kappa(\alpha, \varepsilon)$ such that, for any rational number p/q with $q > 0$, we have*

$$\left|\alpha - \frac{p}{q}\right| > \frac{\kappa(\alpha, \varepsilon)}{q^{2+\varepsilon}}.$$

From the previous argument, we deduce the following.

Corollary 1 (A. Thue) *Let $F \in \mathbb{Z}[X, Y]$ be an irreducible binary form of degree $d \geq 3$. Let $m \in \mathbb{Z}$. Then there are only finitely many (x, y) in $\mathbb{Z} \times \mathbb{Z}$ such that $F(x, y) = m$.*

One main drawback of this argument is that the proof following the original approach by Thue does not produce an effective value for the constant $\kappa(\alpha, \varepsilon)$ when ε is less than $d - 2$. As a consequence, Corollary 1 is not effective; as a matter of fact, upper bounds for the number of solutions (x, y) to the Diophantine equation $F(x, y) = m$ (qualitative statements) can be derived from the proof, but no upper bound for $\max\{|x|, |y|\}$ can be obtained (quantitative statements). We will discuss below (see Sect. 2) another approach which has been suggested by A. O. Gel'fond and worked out by A. Baker, involving lower bounds for linear forms in logarithms and it is effective.

An illuminating presentation of Thue's method is given by D. W. Masser in [27], Chap. 12, where he starts with $x^3 - 2y^3 = m$ and goes on by explaining some of the main ideas behind Thue's proof, building upon Newton's method.

1.4 An Example: $x^3 - 2y^3 = m$

Let us consider the special case of the cubic Thue equation $x^3 - 2y^3 = m$ (see [35]). Let ψ be a positive real function which satisfies

$$\left|\sqrt[3]{2} - \frac{p}{q}\right| > \frac{\psi(q)}{q^3}$$

for each $q > 0$. Let $m \in \mathbb{Z}, m \neq 0$ and let $(x, y) \in \mathbb{Z}^2$ satisfy $x^3 - 2y^3 = m$. Assume x and y are positive (this is no loss of generality). Write

$$x^3 - 2y^3 = (x - \sqrt[3]{2}y)(x^2 + \sqrt[3]{2}xy + \sqrt[3]{4}y^2)$$

and observe that $x^2 + \sqrt[3]{2}xy + \sqrt[3]{4}y^2 > y^2$. We deduce that any solution (x, y) in positive integers of the equation $x^3 - 2y^3 = m$ satisfies

$$\psi(y) \le |m|. \tag{3}$$

If $\psi(q)$ tends to infinity with q, then we get an upper bound for y while x is bounded by

$$x \le \sqrt[3]{2} \max\{\sqrt[3]{|m|}, \sqrt[3]{2}y\}.$$

In the other direction, let ψ be a positive real function such that any solution (x, y) in positive rational integers of $x^3 - 2y^3 = m$ satisfies

$$\psi(y) \le |m|.$$

We write

$$|p^3 - 2q^3| = \left|\sqrt[3]{2} - \frac{p}{q}\right| (p^2 + \sqrt[3]{2}pq + \sqrt[3]{4}q^2)q.$$

If $p \le (3/2)q$, we have $p^2 + \sqrt[3]{2}pq + \sqrt[3]{4}q^2 \le 6q^2$ and we deduce

$$\left|\sqrt[3]{2} - \frac{p}{q}\right| \ge \frac{\psi(q)}{6q^3}.$$

If $p > (3/2)q$, then we have the sharper estimate

$$\left|\sqrt[3]{2} - \frac{p}{q}\right| > \frac{3}{2} - \sqrt[3]{2} > \frac{1}{5}.$$

Liouville's estimate

$$\left|\sqrt[3]{2} - \frac{p}{q}\right| > \frac{1}{6q^3} \tag{4}$$

follows by taking for ψ the constant function $\psi(y) = 1$, while any upper bound for the solutions (x, y) of $x^3 - 2y^3 = m$ implies the validity of (3) with a function $\psi(y)$ tending to infinity with y, and this yields an improvement on Liouville's estimate (4).

In this direction, the sharpest known estimates are due to Bennett [4]:

$$\left|\sqrt[3]{2} - \frac{p}{q}\right| > \frac{1}{4q^{2.5}} \quad \text{and} \quad |x^3 - 2y^3| \geq \sqrt{x}.$$

See also [5] for similar results concerning other algebraic numbers than $\sqrt[3]{2}$.

2 Solving Thue Equation Using Baker's Method

2.1 References

Here is a selection of books having a section devoted to Baker's method for solving Thue equations.

• Baker [2] Chap. 4 (Diophantine equations) Sect. 2. The Thue equation—proof using lower bounds for linear forms in logarithms.

• Shorey et al. [33] Main results arising from Baker's method in 1977, with proofs using lower bounds for linear forms in logarithms.

• Shorey and Tijdeman [32] Chap. 5 (The Thue equation) proof using lower bounds for linear forms in logarithms. Includes effective estimates.

• Sprindžuk [34] Chap. IV (The Thue Equation).

• Baker and Wüstholz [3] Chap. 3 (Diophantine problems) Sect. 3 The Thue equation—sketch of proof using lower bounds for linear forms in logarithms.

• Győry [14] Survey of some important applications of Baker's theory of linear forms in logarithms to Diophantine equations.

• Evertse and Győry [10] Chap. 9 (Decomposable forms equations) Sect. 9.6 (Effective results) Sect. 9.6.1 (Thue equations) Explicit results.

• Bugeaud [7] Chap. 4 Sect. 3 (The Thue equation).

2.2 Thue Equation and Siegel's Unit Equation

We explain some of the basic ideas behind the reduction of the Diophantine equation $F(x, y) = m$ to Siegel's unit equation.

Assume for simplicity $a_0 = 1$, $a_d = \pm 1$, $m = 1$ so that the polynomial f can be written as

$$f(T) = T^d + a_1 T^{d-1} + \cdots + a_{d-1} T \pm 1 = (T - \alpha_1) \ldots (T - \alpha_d)$$

where $\alpha_1, \ldots, \alpha_d$ are units of degree d. The numbers $\beta_i = x - \alpha_i y$ $(i = 1, \ldots, d)$ are also units of degree d since they are algebraic integers satisfying $\beta_1 \cdots \beta_d = 1$. If i_1, i_2, i_3 are distinct indices in $\{1, \ldots, d\}$ (recall $d \geq 3$), eliminating x and y among the three relations

$$\beta_{i_1} = x - \alpha_{i_1} y, \quad \beta_{i_2} = x - \alpha_{i_2} y, \quad \beta_{i_3} = x - \alpha_{i_3} y$$

shows that the determinant

$$\begin{vmatrix} 1 & -\alpha_{i_1} & \beta_{i_1} \\ 1 & -\alpha_{i_2} & \beta_{i_2} \\ 1 & -\alpha_{i_3} & \beta_{i_3} \end{vmatrix}$$

is 0. This yields the so-called *Siegel unit equation*

$$\beta_{i_1}(\alpha_{i_2} - \alpha_{i_3}) + \beta_{i_2}(\alpha_{i_3} - \alpha_{i_1}) + \beta_{i_3}(\alpha_{i_1} - \alpha_{i_2}) = 0.$$

The main result on Siegel's unit equation is that given γ_1, γ_2 in $K^\times$, the set of pairs (u_1, u_2) of units in a number field K satisfying $\gamma_1 u_1 + \gamma_2 u_2 = 1$ is finite. In homogeneous form, the result is the following: if $\gamma_1, \gamma_2, \gamma_3$ are in $K^\times$, if we consider the equation $\gamma_1 u_1 + \gamma_2 u_2 + \gamma_3 u_3 = 0$, then the set of $(u_1/u_3, u_2/u_3)$ is finite. Baker's method gives an effective upper bound for the heights of the solutions.

Once we know that the set of numbers

$$\frac{\beta_{i_1}(\alpha_{i_2} - \alpha_{i_3})}{\beta_{i_2}(\alpha_{i_3} - \alpha_{i_1})},$$

for (x, y) solution of $F(x, y) = m$, is finite, we deduce that the set of quotients β_{i_1}/β_{i_2} is finite, hence x/y belongs to a finite set E; if $y = vx$ with $v \in E$, then $F(x, y) = x^d F(1, v)$ and the equation $x^d F(1, v) = m$ yields the desired upper bound for $|x|$.

2.3 Lower Bounds for Linear Forms in Logarithms and Siegel's Unit Equation

Let $\alpha_1, \dots, \alpha_n$ be nonzero algebraic numbers and $b_1, \dots, b_n$ be rational integers such that

$$\alpha_1^{b_1} \cdots \alpha_n^{b_n} \neq 1.$$

Define $B = \max\{2, |b_1|, \dots, |b_n|\}$. A "trivial" estimate "à la Liouville" is

$$|\alpha_1^{b_1} \dots \alpha_n^{b_n} - 1| \geq e^{-C_1 B},$$

where C_1 is an explicit constant depending only on $\alpha_1, \dots, \alpha_n$. Methods from transcendental number theory involving the quantity $\beta_1 \log \alpha_1 + \dots + \beta_n \log \alpha_n$ yield the refinement

$$|\alpha_1^{b_1} \dots \alpha_n^{b_n} - 1| \geq B^{-C_2}, \tag{5}$$

where C_2 is also an explicit constant depending only on $\alpha_1, \dots, \alpha_n$. This estimate is optimal as far as the dependence on B is concerned (but optimal estimates for C_2 have not yet been achieved).

Let γ_1, γ_2 be nonzero elements in a number field K. Let $\varepsilon_1, \dots, \varepsilon_r$ be a basis of the group of units of K. Let (u_1, u_2) be two units in K such that

$$\gamma_1 u_1 + \gamma_2 u_2 = 1.$$

We write

$$\frac{u_1}{u_2} = \zeta \varepsilon_1^{b_1} \cdots \varepsilon_r^{b_r}$$

where ζ is a root of unity in K and $b_1, \dots, b_r$ are rational integers. Set

$$\gamma_0 = \frac{-\gamma_1 \zeta}{\gamma_2}.$$

We use the fundamental Diophantine estimate (5) to obtain a lower bound for the modulus of any complex conjugate of the left-hand side of

$$\gamma_0 \varepsilon_1^{b_1} \cdots \varepsilon_r^{b_r} - 1 = \frac{-1}{\gamma_2 u_2}.$$

An auxiliary lemma (e.g., Lemma 5.1 of [32] or Lemma 5 of [25]) shows that one can choose such a complex embedding for which the modulus of the right-hand side is small, so that we end up with an upper bound for the heights of u_1 and u_2.

Essentially, the three statements:

- finiteness of the set of solutions to Thue equations,
- finiteness of the set of solutions of the unit equation, and
- nontrivial refinement of Liouville's Theorem

are equivalent. An upper bound for the number of exceptional solutions for one of these statements implies such an upper bound for the two others; an effective result on one of them (upper bound for the exceptional solutions) yields an effective result on the two others. See [18, 35].

An effective result on the Thue equation is the following: let $F \in \mathbb{Z}[X, Y]$ be an irreducible binary form of degree ≥ 3; let $(x, y) \in \mathbb{Z}^2$ and let $m = F(x, y)$. Then

$$\max\{|x|, |y|\} \leq m^{\kappa},$$

where κ is a positive effective absolute constant depending only on F; explicit formulae are available (see for instance [10]). At the early stages of Baker's method, such constants were huge; drastic improvements have been achieved; nowadays these estimates are good enough for solving explicitly Thue equations with coefficients which are not too large. Algorithms using this approach are implemented in computation packages.

3 Families of Thue Equations

3.1 Historical Survey

Given a family $F_t(X, Y)$, $t \in I$ of binary forms of degree ≥ 3, the first goal is to prove, under suitable assumptions, that for all $m > 0$ there are only finitely many $(t, x, y) \in I \times \mathbb{Z} \times \mathbb{Z}$ satisfying $F_t(x, y) = m$. Sometimes some subsets of (t, x, y), corresponding to "trivial solutions", are excluded. The second goal is to give an upper bound for the exceptional solutions.

A survey on these question is [15]. Further results can be found in [10]. See also [1] for another approach (the family $x^3 - (t^3 - 1)y^3 = 1$ is quoted).

3.2 Idea of the Proof

Let $f \in \mathbb{Z}[T]$ be an irreducible polynomial of degree $d \geq 3$ and let

$$F(X, Y) = Y^d f(X/Y) \in \mathbb{Z}[X, Y].$$

Denote by α a root of f, by K the field $\mathbb{Q}(\alpha)$, and by υ a unit in K of infinite order. For $a \in \mathbb{Z}$ we denote by f_a the irreducible polynomial of $\alpha\upsilon^a$. We assume that f_a

has degree d. Let F_a be the binary form $Y^d f_a(X/Y)$ so that $f_0 = f$ and $F_0 = F$. Given $m \in \mathbb{Z}$, $m \neq 0$, we consider the set of triples (x, y, a) in $\mathbb{Z}^3$ such that

$$F_a(x, y) = m.$$

Let $\sigma_1, \ldots, \sigma_d$ be the embeddings of K in $\mathbb{C}$. We define

$$\alpha_i = \sigma_i(\alpha), \quad \upsilon_i = \sigma_i(\upsilon) \qquad (i = 1, \ldots, d),$$

so that

$$f_a(T) = a_0 \prod_{i=1}^{d} (T - \alpha_i \upsilon_i^a),$$

$$F_a(X, Y) = Y^d f_a(X/Y) = a_0 \prod_{i=1}^{d} \left(X - \alpha_i \upsilon_i^a Y\right).$$

Let m be a nonzero integer and (x, y, a) be a solution of $F_a(x, y) = m$ with $\mathbb{Q}(\alpha\upsilon^a) = K$. For $i = 1, \ldots, d$, set $\beta_i = x - \alpha_i \upsilon_i^a y$. We have

$$a_0 \beta_1 \cdots \beta_d = m.$$

Eliminating x and y among three equations $\beta_i = x - \alpha_i \upsilon_i^a y$ for $i = i_1, i_2, i_3$ yields the unit equation

$$\beta_{i_1} \alpha_{i_2} \upsilon_{i_2}^a - \beta_{i_1} \alpha_{i_3} \upsilon_{i_3}^a + \beta_{i_2} \alpha_{i_3} \upsilon_{i_3}^a - \beta_{i_2} \alpha_{i_1} \upsilon_{i_1}^a + \beta_{i_3} \alpha_{i_1} \upsilon_{i_1}^a - \beta_{i_3} \alpha_{i_2} \upsilon_{i_2}^a = 0.$$

A first approach is to use Schmidt's Subspace Theorem and its consequence on the generalized S-unit equations: given a finite set of places S of a number field K and an integer n, the set of nondegenerate solutions $(u_1, \ldots, u_n)$ in S-units of the equation

$$u_1 + \cdots + u_n = 1$$

is finite. *Nondegenerate* means that no nontrivial subsum of $u_1 + \cdots + u_n$ vanishes. A technical difficulty is that we need to deal with degenerate solutions. This approach yields strong general but ineffective results [19].

Another approach, which is effective, is to use Baker's method. This is efficient as soon as two of the six terms on the left-hand side have a size which is much larger than the sum of all other four terms (besides, these two terms should not yield a zero subsum). So far, this has been achieved only in special cases [21–26].

3.3 Joint Papers with Claude Levesque

We give here a summary of the results in a sequence of joint papers with C. Levesque, which was initiated during our visit to IMPA in Rio de Janeiro in 2010. The initial goal, which was to solve the family of equations obtained from Thomas equations (see [15]) by including powers of units, has been achieved in [24].

In [18], we work out equivalence statements between assertions dealing with several Diophantine questions: Thue–Mahler equations, S-unit equation, integral points on $\mathbb{P}^1(K)$ minus three points.

Our first results [8, 19, 20] were based on Schmidt's Subspace Theorem and therefore were not effective—but they were very general. We obtained families of Thue–Mahler equations having only finitely many solutions and we gave upper bounds for the number of solutions; but we were not able to give upper bounds for the solutions themselves, hence we could not solve the equations.

Our first main new results were proved in [19]. Some consequences on Diophantine approximation were given in [20].

Here is a special case of Corollary 3.6 of [19] which deals with Thue–Mahler equations.

Theorem 3 *Let K be a number field and Γ a finitely generated subgroup of $K^\times$. For $\gamma \in \Gamma$, denote by $f_\gamma \in \mathbb{Z}[X]$ the irreducible polynomial of γ and by $F_\gamma \in \mathbb{Z}[X, Y]$ its homogeneous version. Then the set of $\gamma \in \Gamma$ satisfying $[\mathbb{Q}(\gamma) : \mathbb{Q}] \geq 3$ for which there exists $(x, y) \in \mathbb{Z}^2$ with $F_\gamma(x, y) \in \Gamma$ and $xy \neq 0$ is finite.*

Proof Since K has only finitely many subfields, it suffices to prove that the set of $\gamma \in \Gamma$ satisfying $\mathbb{Q}(\gamma) = K$ for which there exists $(x, y) \in \mathbb{Z}^2$ with $F_\gamma(x, y) \in \Gamma$ and $xy \neq 0$ is finite.

Assume $\gamma \in \Gamma$ satisfies $\mathbb{Q}(\gamma) = K$ and assume that there exists $(x, y) \in \mathbb{Z}^2$ with $F_\gamma(x, y) \in \Gamma$ and $xy \neq 0$. Let $\alpha \in K$ with $K = \mathbb{Q}(\alpha)$. Let S be a finite set of places of K such that $\alpha \in O_S^\times$ et $\Gamma \subset O_S^\times$. Corollary 3.6 of [19] with $t = 0, \varepsilon = \gamma/\alpha$ yields the result. □

By taking for Γ the group of units $\mathbb{Z}_K^\times$ in K, we deduce the following result.

Corollary 2 *Let K be a number field. The set of units $\varepsilon \in \mathbb{Z}_K^\times$ of degree ≥ 3 for which there exists $(x, y) \in \mathbb{Z}^2$ with $F_\varepsilon(x, y) = \pm 1$ and $xy \neq 0$ is finite.*

Another result from [19] is the following. *Let $S = \{p_1, \ldots, p_s\}$ be a finite set of prime numbers, $f \in \mathbb{Z}[X]$ an irreducible polynomial of degree $d \geq 3$, α a root of f, K the number field $\mathbb{Q}(\alpha)$, $\sigma_1, \ldots, \sigma_d$ the embeddings of K into $\mathbb{C}$. For each S-unit $\varepsilon \in O_S^\times$, define $F_\varepsilon(X, Y) \in \mathbb{Z}[X, Y]$ by*

$$F_\varepsilon(X, Y) = a_0\big(X - \sigma_1(\alpha\varepsilon)Y\big)\big(X - \sigma_2(\alpha\varepsilon)Y\big)\ldots\big(X - \sigma_d(\alpha\varepsilon)Y\big).$$

Let $m \in \mathbb{Z} \setminus \{0\}$. Then the set of $(x, y, \varepsilon, z_1, \ldots, z_s)$ in $\mathbb{Z}^2 \times O_S^\times \times \mathbb{N}^s$ satisfying

$$F_\varepsilon(x, y) = mp_1^{z_1} \dots p_s^{z_s}$$

with $xy \neq 0$, $\gcd(xy, p_1 \dots p_s) = 1$, *and* $[\mathbb{Q}(\alpha\varepsilon) : \mathbb{Q}] \geq 3$, *is finite.*

Theorem 3 implies finiteness results for families of Thue–Mahler equations. It is not effective: upper bounds for the number of solutions could be deduced, but not upper bounds for the heights of the solutions.

In [8], which is a joint paper involving also Yann Bugeaud, we obtained an upper bound for the number of solutions of simultaneous Brahmagupta–Fermat–Pell–Mahler equations: given rational integers $a_1, b_1, c_1, a_2, b_2, c_2$ and prime numbers $p_1, \dots, p_s$, we considered the system of equations

$$\begin{cases} a_1X^2 + b_1XZ + c_1Z^2 = \pm p_1^{m_1} \dots p_s^{m_s}, \\ a_2Y^2 + b_2YZ + c_2Z^2 = \pm p_1^{n_1} \dots p_s^{n_s}, \end{cases}$$

where the unknowns $x, y, z, m_1, \dots, m_s, n_1, \dots, n_s$ take their values in the set of rational integers with $m_1, \dots, m_s, n_1, \dots, n_s$ nonnegative.

Our more recent papers provide effective results for families of Thue equations by means of Baker's method. One main goal is to prove the following conjecture.

Conjecture 1 *Let α be an algebraic number of degree $d \geq 3$ over $\mathbb{Q}$. We denote by K, the algebraic number field $\mathbb{Q}(\alpha)$, by $f \in \mathbb{Z}[X]$ the irreducible polynomial of α over $\mathbb{Z}$, by $\mathbb{Z}_K^\times$ the group of units of K, and by r the rank of the abelian group $\mathbb{Z}_K^\times$. For any unit $\varepsilon \in \mathbb{Z}_K^\times$ such that the degree $\delta = [\mathbb{Q}(\alpha\varepsilon) : \mathbb{Q}]$ is ≥ 3, we denote by $f_\varepsilon(X) \in \mathbb{Z}[X]$ the irreducible polynomial of $\alpha\varepsilon$ over $\mathbb{Z}$ (uniquely defined upon requiring that the leading coefficient be > 0) and by F_ε the irreducible binary form defined by $F_\varepsilon(X, Y) = Y^\delta f_\varepsilon(X/Y) \in \mathbb{Z}[X, Y]$. Then there exists an effectively computable constant $\kappa > 0$, depending only upon α, such that, for any $m \geq 2$, each solution $(x, y, \varepsilon) \in \mathbb{Z}^2 \times \mathbb{Z}_K^\times$ of the inequation $|F_\varepsilon(x, y)| \leq m$ with $xy \neq 0$ and $[\mathbb{Q}(\alpha\varepsilon) : \mathbb{Q}] \geq 3$ satisfies*

$$\max\{|x|, |y|, e^{\mathrm{h}(\alpha\varepsilon)}\} \leq m^\kappa.$$

In [21], we prove Conjecture 1 when the field K is a non-totally real cubic field. In [25], we prove Conjecture 1 in the more general case where the field K has at most one real embedding. In [22], we prove Conjecture 1 when one requests the unknown ε to belong to a subset of the group of units of K; we show that this subset contains a positive proportion of all units as soon as the degree of K is at least 4.

The papers [21, 23, 24, 26] deal with the special case where one restricts to a rank one subgroup of the group of units, namely when $\varepsilon = \upsilon^a$ with $a \in \mathbb{Z}$.

The main result of [21], which deals only with non-totally real cubic equations, is a special case of the main result of [26]; the "constants" in [21] depend on α and υ while in [26] they depend only on the degree d. The main result of [22] deals with Thue equations twisted by a set of units which is not supposed to be a group of rank one, but it involves an assumption (namely that at least two of the conjugates of υ have a modulus as large as a positive power of $\lceil\upsilon\rceil$) which we do not need in [26]. Our Theorem in [26] also improves the main result of [23]: we remove the assumption that the unit is totally real (besides, the result of [23] is not explicit in terms of the heights and regulator). We also notice that part (iii) of Theorem 1.1 of [24] follows from our Theorem in [26]. The main result of [25] does not assume that the twists are done by a group of units of rank one, but it needs a strong assumption which does not occur in [26], namely that the field K has at most one real embedding.

A very recent joint work with E. Fouvry and C. Levesque, already quoted in Sect. 1.2, deals with the family of cyclotomic binary forms [12]. One motivation came from the fact that in [26], we needed an assumption that some number was not a root of unity. It was a natural task to study the special case of roots of unity, which gives rise to the sequence of cyclotomic binary forms.

4 A Guide to Further References

One of the main references is the classical paper of C. L. Siegel in 1929, which has been recently translated into English. The reference [38] includes the English translation *On some applications of Diophantine approximations* by Clemens Fuchs, of the original text by C. L. Siegel in German *Über einige Anwendungen diophantischer Approximationen*, which is also reproduced in [38], together with comments by Clemens Fuchs and Umberto Zannier *Integral points on curves: Siegel's theorem after Siegel's proof.*

Another reference is Zannier [37] Chap. 2 (Thue's equations and rational approximations), where full proofs are given with lots of supplements.

There are many references on Diophantine geometry and Schmidt Subspace Theorem, including the following ones:

- Lang [17], Chap. 7 (The Thue–Siegel–Roth Theorem).
- Serre [31], Chap. 7 (Siegel's method), Chap. 8 (Baker's method).
- Schmidt [30], Chap. III (The Thue equation); also Chap. V (Diophantine equations in More Than Two Variables).
- Zannier [36], Chap. 1 (Diophantine Approximation and Diophantine equations) Sect. 1.2 (From Thue to Roth); also Chap. II (Schmidt's Subspace Theorem and S-unit equations) and Chap. III (Integral points on curves and other varieties).
- Bugeaud [6], Chap. 2 (Approximation to algebraic numbers), Sect. 2.1 (Rational approximation), Sect. 2.2 (Effective rational approximation).
- Hu and Yang [16], Chap. 6 (Roth Theorem); also Chap. 7 (Subspace Theorem) and Chap. 8 (Vojta's conjectures)
- Corvaja [9], Chap. 3 (The theorems of Thue and Siegel); also includes results on Hilbert Irreducibility Theorem and integral points on surfaces.

References

1. F. Amoroso, D. Masser, U. Zannier, Bounded height in pencils of finitely generated subgroups. Duke Math. J. **166**, 2599–2642 (2017), arXiv:1509.04963 [math.NT]
2. A. Baker, *Transcendental Number Theory*, Cambridge Mathematical Library (1975), 2nd edn (1990)
3. A. Baker, G. Wüstholz, *Logarithmic Forms and Diophantine Geometry*, vol. 9. New Mathematical Monographs (Cambridge University Press , Cambridge, 2007)
4. M.A. Bennett, Effective measures of irrationality for certain algebraic numbers. J. Austral. Math. Soc. Ser. A **62**, 329–344 (1997)
5. M.A. Bennett, Explicit lower bounds for rational approximation to algebraic numbers. Proc. London Math. Soc. **75**(3), 63–78 (1997)
6. Y. Bugeaud, *Approximation by Algebraic Numbers*, vol. 160. Cambridge Tracts in Mathematics (Cambridge University Press, Cambridge, 2004)
7. Y. Bugeaud, *Linear Forms in Logarithms and Applications, IRMA Lectures in Mathematics and Theoretical Physics*, vol. 28. European Mathematical Society (2018). http://www.ems-ph.org/books/book.php?proj_nr=228
8. Y. Bugeaud, C. Levesque, M. Waldschmidt, Équations de Fermat-Pell–Mahler simultanées. Publ. Math. Debrecen. **79**(3–4), 357–366 (2011)
9. P. Corvaja, *Integral Points on Algebraic Varieties*, vol. 3. Institute of Mathematical Sciences Lecture Notes (Hindustan Book Agency, New Delhi, 2016)
10. J.-H. Evertse, K. Győry, *Unit Equations in Diophantine Number Theory*, vol. 146. Cambridge Studies in Advanced Mathematics (Cambridge University Press, Cambridge, 2015)
11. A. Faisant, *L'équation diophantienne du second degré*, vol. 1430 of Actualités Scientifiques et Industrielles [Current Scientific and Industrial Topics], Hermann, Paris, 1991. Collection Formation des Enseignants et Formation Continue. [Collection on Teacher Education and Continuing Education]
12. E. Fouvry, C. Levesque, M. Waldschmidt, *Representation of integers by cyclotomic binary forms*. Acta Arithmetica **184.1**, 67–86 (2018). arXiv:1712.09019 [math.NT]
13. K. Győry, Représentation des nombres entiers par des formes binaires. Publ. Math. Debrecen **24**(3–4), 363–375 (1977)
14. K. Győry, Solving Diophantine, equations by Baker's theory, in *A Panorama of Number Theory or the View from Baker's Garden (Zürich* (Cambridge University Press, Cambridge, 2002), pp. 38–72
15. C. Heuberger, Parametrized Thue equations –a survey, in *Proceedings of the RIMS Symposium "Analytic Number Theory and Surrounding Areas*, RIMS Kôkyûroku 2004, vol. 1511 (Kyoto, 2006), pp. 82–91
16. P.-C. Hu, C.-C. Yang, *Distribution Theory of Algebraic Numbers*, vol. 45. De Gruyter Expositions in Mathematics (2008)
17. S. Lang, *Fundamentals of Diophantine geometry* (Springer, New York, 1983)
18. C. Levesque, M. Waldschmidt, Some remarks on diophantine equations and diophantine approximation. Vietnam J. Math. **39**3, 343–368 (2011). arXiv:1312.7200 [math.NT]
19. C. Levesque, M. Waldschmidt, Familles d'équations de Thue–Mahler n'ayant que des solutions triviales. Acta Arith. **155**, 117–138 (2012). arXiv:1312.7202 [math.NT]
20. C. Levesque, M. Waldschmidt, Approximation of an algebraic number by products of rational numbers and units. J. Aust. Math. Soc. **93**(1–2), 121–131 (2013). arXiv:1312.7203 [math.NT]
21. C. Levesque, M. Waldschmidt, Families of cubic Thue equations with effective bounds for the solutions, Number Theory and Related Fields, in *Memory of Alf van der Poorten, Springer Proceedings in Mathematics and Statistics*, vol. 43, ed by J.M. Borwein et al., pp. 229–243 (2013). arXiv:1312.7204 [math.NT]
22. C. Levesque, M. Waldschmidt, Solving effectively some families of Thue Diophantine equations. Moscow J. Comb. Number Theory **3**(3–4), 118–144 (2013). arXiv:1312.7205 [math.NT]

23. C. Levesque, M. Waldschmidt, Familles d'équations de Thue associées à un sous-groupe de rang 1 d'unités totalement réelles d'un corps de nombres, in *SCHOLAR—a Scientific Celebration Highlighting Open Lines of Arithmetic Research, 2013 (volume dedicated to Ram Murty), CRM collection Contemporary Mathematics", AMS*, vol. 655, 117–134 (2015). http://www.ams.org/books/conm/655/, arXiv: 1505.06656 [math.NT]
24. C. Levesque, M. Waldschmidt, A family of Thue equations involving powers of units of the simplest cubic fields. J. Théor. Nombres Bordx. **27**(2), 537–563 (2015). arXiv:1505.06708 [math.NT]
25. C. Levesque, M. Waldschmidt, Solving simultaneously Thue Diophantine equations: almost totally imaginary case, in *Ramanujan Mathematical Society*, Lecture Notes Series, vol. 23, 137–156 (2016). arXiv: 1505.06653 [math.NT]
26. C. Levesque, M. Waldschmidt, Families of Thue equations associated with a rank one subgroup of the unit group of a number field. Mathematika **63**(3), 1060–1080 (2017). arXiv: 1701.01230 [math.NT]
27. D. Masser, *Auxiliary Polynomials in Number Theory*, vol. 207. Cambridge Tracts in Mathematics (Cambridge University Press, Cambridge, 2016)
28. L.J. Mordell, *Diophantine Equations, Pure and Applied Mathematics*, vol. 30 (Academic, New York, 1969)
29. N.J. Sloane, *The On–line Encyclopedia of Integer Sequences*. https://oeis.org/
30. W.M. Schmidt, *Diophantine Approximations and Diophantine Equations*, vol. 1467. Lecture Notes in Mathematics (Springer, Berlin, 1991)
31. J.-P. Serre, *Lectures on the Mordell-Weil theorem*, 3rd edn. 1997. Aspects of Mathematics (Friedr. Vieweg and Sohn, Braunschweig, 1989)
32. T.N. Shorey, R. Tijdeman, *Exponential Diophantine Equations*, vol. 87. Cambridge Tracts in Mathematics (Cambridge University Press, Cambridge, 1986)
33. T.N. Shorey, A.J. Van der Poorten, R. Tijdeman, A. Schinzel, Applications of the Gel'fond-Baker method to Diophantine equations, in *Transcendence Theory: Advances and Applications* (Cambridge, 1977), pp. 59–77
34. V.G. Sprindžuk, *Classical Diophantine Equations*, vol. 1559. Lecture Notes in Mathematics (Springer, Berlin, 1993). Translated from the 1982 Russian original
35. M. Waldschmidt, Diophantine equations and transcendental methods (written by Noriko Hirata), in *Transcendental Numbers and Related Topics, RIMS Kôkyûroku*, vol. 599. n°8 (Kyoto, 1986), pp. 82-94. http://www.kurims.kyoto-u.ac.jp/~kyodo/kokyuroku/contents/599.html
36. U. Zannier, *Some Applications of Diophantine Approximation to Diophantine Equations with Special Emphasis on the Schmidt Subspace Theorem* (Forum Editrice Universitaria Udine srl, Udine, 2003)
37. U. Zannier, *Lecture Notes on Diophantine Analysis*, vol. 8. Appunti. Scuola Normale Superiore di Pisa (Nuova Serie) (Edizioni della Normale, Pisa, 2009)
38. U. Zannier, ed., *On Some Applications of Diophantine Approximations*, vol. 2. Quaderni/Monographs (Edizioni della Normale, Pisa, 2014). A translation of Carl Ludwig Siegel's "Über einige Anwendungen diophantischer Approximationen" by Clemens Fuchs, With a commentary and the article "Integral points on curves: Siegel's theorem after Siegel's proof" by Fuchs and Umberto Zannier

A Lower Bound for the Class Number of Certain Real Quadratic Fields

Fuminori Kawamoto and Yasuhiro Kishi

1 Introduction

For a sequence $A = \langle a_1, a_2, \ldots, a_m \rangle$ of $m(\geq 2)$ positive integers, we define q_n and $r_n (0 \leq n \leq m+1)$ by

$$\begin{cases} q_0 = 0, & q_1 = 1, \quad q_n = a_{n-1}q_{n-1} + q_{n-2} \ (2 \leq n \leq m+1), \\ r_0 = 1, & r_1 = 0, \quad r_n = a_{n-1}r_{n-1} + r_{n-2} \ (2 \leq n \leq m+1), \end{cases}$$

inductively. If either $r_{m+1} = 2q_m$ or $r_{m+1} = 2q_m - q_{m+1}$ holds, we say that A is of *pre-ELE type with length* m. Specially A is said to be of *pre-ELE*$_1$ *type* (resp. *pre-ELE*$_2$ *type) with length* m if $r_{m+1} = 2q_m$ (resp. $r_{m+1} = 2q_m - q_{m+1}$) holds. Put

$$\varepsilon^A := \begin{cases} 0 & \text{if } A : \text{pre-ELE}_1 \text{ type}, \\ 1 & \text{if } A : \text{pre-ELE}_2 \text{ type}. \end{cases}$$

Let b be a positive integer such that

$$b \geq 2, \ 2b > a_1, \ldots, a_m \ \textit{and} \ b \equiv (-1)^m q_m r_m \pmod{q_{m+1}} \tag{1.1}$$

$$(\textit{resp.}\ b \geq 4, \ 2b + 2 > a_1, \ldots, a_m \ \textit{and} \ b \equiv (-1)^m q_m (q_m + r_m) \pmod{q_{m+1}}), \tag{1.2}$$

F. Kawamoto
Faculty of Science, Department of Mathematics, Gakushuin University,
1-5-1 Mejiro, Toshima-ku, Tokyo 171-8588, Japan
e-mail: fuminori.kawamoto@gakushuin.ac.jp

Y. Kishi (✉)
Faculty of Education, Department of Mathematics, Aichi University of Education,
1 Hirosawa Igaya-cho, Kariya-shi, Aichi 448-8542, Japan
e-mail: ykishi@auecc.aichi-edu.ac.jp

K. Chakraborty et al. (eds.), *Class Groups of Number Fields and Related Topics*,
https://doi.org/10.1007/978-981-15-1514-9_4

if A is of pre-ELE$_1$ type (resp. pre-ELE$_2$ type) with length m. Then the sequence A, b of $m+1$ positive integers is said to be of *ELE$_1$ type* (resp. *ELE$_2$ type*) (cf. [1, Proposition 4.1]). Moreover, from [1, Theorem 2], we have the following.

Proposition 1.1 *Let the notation be as above.*

(1) *Define d by*

$$d := (b+\varepsilon^A)^2 + \frac{2(br_{m+1}+r_m)+\varepsilon^A r_{m+1}}{q_{m+1}}.$$

Then d is a positive integer with $[\sqrt{d}] = b + \varepsilon^A$ *and*

$$d \equiv \begin{cases} 2 \pmod 4 & \text{if } b \text{ is even,} \\ 3 \pmod 4 & \text{if } b \text{ is odd.} \end{cases}$$

Furthermore, the simple continued fraction expansion of $\sqrt{d}$ *is*

$$\sqrt{d} = [b+\varepsilon^A, \overline{a_1, \ldots, a_m, b, a_m, \ldots, a_1, 2b+2\varepsilon^A}]$$

with minimal period $2m+2$.

(2) *Let d be as in* (1) *and* m_d *the Yokoi invariant of d* (*cf.* [1, Remark 1.4 (2)]). *Then we have* $m_d = 2q_{m+1}^2$ *if m is odd, and* $m_d = 2q_{m+1}^2 - 1$ *if m is even.*

Remark 1.1 (1) Let d be a non-square positive integer. If the simple continued fraction expansion of $\sqrt{d}$ has even period, then it is of the form $\sqrt{d} = [a_0, \overline{a_1, \ldots, a_m, a_{m+1}, a_m, \ldots, a_1, 2a_0}]$. Then we say that the sequence $a_1, a_2, \ldots, a_m, a_{m+1}$ is the *primary symmetric part* of the simple continued fraction expansion of $\sqrt{d}$. Proposition 1.1 gives a way of constructing every positive integer d of minimal type such that the primary symmetric part of the simple continued fraction expansion of $\sqrt{d}$ with even period (≥ 6) is of ELE type. (As for the definition of "minimal type", see [3, Definition 3.1].)

(2) For any prime p with $p \equiv 3 \pmod 4$, the primary symmetric part of the simple continued fraction expansion of $\sqrt{p}$ is of ELE type if its minimal period is greater than or equal to 6 (cf. [1, Corollary 1, Theorem 1]).

Now let b^A denote the smallest positive integer b satisfying the three conditions (1.1) (resp. (1.2)), if A is of pre-ELE$_1$ type (resp. pre-ELE$_2$ type) with length m, and put

$$b(t) := b^A + (t-1)q_{m+1}$$

for each positive integer $t \geq 1$. Then the sequence $A, b(t)$ satisfies the condition (1.1) (resp. (1.2)) again, and so $A, b(t)$ is of ELE$_1$ type (resp. ELE$_2$ type). Then we can define a positive integer $d(t)$ as in Proposition 1.1 such that the simple continued fraction expansion of $\sqrt{d(t)}$ is

$$\sqrt{d(t)} = [b(t) + \varepsilon^A, \overline{a_1, \ldots, a_m, b(t), a_m, \ldots, a_1, 2b(t) + 2\varepsilon^A}]$$

with minimal period $2m + 2$. By a straightforward calculation, we have

$$d(t) = q_{m+1}^2(t-1)^2 + 2(q_{m+1}b^A + 2q_m)(t-1) + d(1),$$

where

$$d(1) = (b^A + \varepsilon^A)^2 + \frac{2(b^A r_{m+1} + r_m) + \varepsilon^A r_{m+1}}{q_{m+1}}.$$

In particular, the sequence $\{d(t)\}_{t\geq 1}$ is strictly monotonously increasing ([2, Proposition 2.1 (1)]). Moreover, the set $\{d(t) \mid t \in \mathbb{Z}, t \geq 1\}$ contains infinitely many square-free elements ([2, Proposition 2.1 (2)]).

In this paper, we consider the class number of $\mathbb{Q}(\sqrt{d(t)})$ for the case where $d(t)$ is square free. Throughout this paper, for a square-free positive integer $d > 1$, h_d and ε_d denote the class number and the fundamental unit (>1) of the quadratic field $\mathbb{Q}(\sqrt{d})$, respectively.

For brevity, we put

$$J = J^A := b^A + \varepsilon^A = [\sqrt{d(1)}] > 0, \ C := \frac{0.3275}{32}.$$

The following is the main theorem of this paper.

Theorem 1 *Under the above setting, let* $m \geq 4$. *Assume that* $J > e^8$ $(= 2980.9579\cdots)$ *and*

$$\frac{CJ^{7/8} - \log(J+1)}{\log(4J^2/9 + 1)} + 1 \geq m. \tag{1.3}$$

Then for $d \in \{d(t) \mid t \in \mathbb{Z}, t \geq 1, d(t) : \textit{square-free}\}$, *we have* $h_d > 1$ *with one more possible exception.*

This paper is organized as follows. In Sect. 2, we give a lower bound for the class number of $\mathbb{Q}(\sqrt{d(t)})$ by using Tatuzawa's theorem and the Yokoi invariant. In Sect. 3, we prove Theorem 1. In the last section, Sect. 4, we give a family of real quadratic fields whose class numbers are not equal to one, by applying Theorem 1 to a sequence $A = \langle 2, \ldots, 2, 2, 1\rangle$ of pre-ELE$_2$ type with length $m(\geq 4)$.

2 A Lower Bound for the Class Number

Tatuzawa's theorem (Tatuzawa [6, Theorem 2]) reads that for any $s \geq 11.2$, square-free $d \geq e^s$ and with one possible exception of d, it holds

$$h_d \log \varepsilon_d > \frac{0.3275 \cdot d^{(s-2)/2s}}{s}$$

(see, for example, Yokoi [7, p. 187]).

From now on, we consider the case where

$$d(t) = q_{m+1}^2(t-1)^2 + 2(q_{m+1}b^A + 2q_m)(t-1) + d(1)$$

is square free. Substituting $s = 16$ in the above inequality, we have

$$h_d \log \varepsilon_d > \frac{0.3275 \cdot d^{7/16}}{16} \tag{2.1}$$

for any $d \in \{d(t) \mid t \in \mathbb{Z},\ t \geq 1, d(t) : \text{square-free},\ d(t) \geq e^{16}\}$ with one more possible exception. Now let us give an upper bound for $\log \varepsilon_d$ by using the Yokoi invariant.

Lemma 2.1 *Let the notation be as above. For* $d \in \{d(t) \mid t \in \mathbb{Z},\ t \geq 1, d(t) :$ *square-free*$\}$, *we have*

$$\log \varepsilon_d < \log \left(2 \prod_{k=1}^{m} a_k^2 \cdot \prod_{k=2}^{m} \left(1 + \frac{1}{a_{k-1}a_k}\right)^2 + 1 \right) + 2\log(b + \varepsilon^A + 1), \tag{2.2}$$

where $b := b(t)$ *for simplicity.*

Proof Since the minimal period of the simple continued fraction expansion of $\sqrt{d}$ is greater than or equal to 6, we have $d \geq 19 > 13$. Then it holds that

$$m_d d < \varepsilon_d < (m_d + 1)d$$

(see [7, Proof of Theorem 1.1]). Hence we have

$$\log \varepsilon_d < \log(2q_{m+1}^2 + 1) + \log d,$$

because $2q_{m+1}^2 \geq m_d > 0$ holds by Proposition 1.1 (2). Here it holds that

$$b + \varepsilon^A < \sqrt{d} < b + \varepsilon^A + 1. \tag{2.3}$$

This implies that

$$\log d < 2\log(b + \varepsilon^A + 1),$$

and hence, we get

$$\log \varepsilon_d < \log(2q_{m+1}^2 + 1) + 2\log(b + \varepsilon^A + 1). \tag{2.4}$$

On the other hand, it holds in general that

$$q_{m+1}^2 \leq \prod_{k=1}^{m} a_k^2 \cdot \prod_{k=2}^{m} \left(1 + \frac{1}{a_{k-1}a_k}\right)^2$$

(see Lang [4, p. 34]). Thus,

$$\log(2q_{m+1}^2 + 1) \leq \log\left(2\prod_{k=1}^{m} a_k^2 \cdot \prod_{k=2}^{m} \left(1 + \frac{1}{a_{k-1}a_k}\right)^2 + 1\right).$$

From this, together with (2.4), we obtain (2.2). □

Put

$$I = I^A := q_{m+1} > 0, \ J = J^A := b^A + \varepsilon^A > 0, \ M = M^A := \frac{e^8 - J}{I}$$

and

$$L = L^A := \log\left(2\prod_{k=1}^{m} a_k^2 \cdot \prod_{k=2}^{m} \left(1 + \frac{1}{a_{k-1}a_k}\right)^2 + 1\right).$$

For $x(\geq 0) \in \mathbb{R}$, define

$$f(x) := \frac{(Ix + J)^{7/8}}{2\log(Ix + J + 1) + L}.$$

Then the following holds.

Theorem 2 *Under the above setting, for* $d \in \{d(t) \mid t \in \mathbb{Z},\ t \geq \max\{1, [M] + 2\},$ $d(t)$ *: square-free*$\}$, *we have*

$$h_d > \begin{cases} \dfrac{0.3275}{16} f([M] + 1) & \text{if } J < e^8, \\ \dfrac{0.3275}{16} f(0) & \text{if } J > e^8 \end{cases} \tag{2.5}$$

with one more possible exception.

For the proof of Theorem 2, we prepare the following lemma.

Lemma 2.2 *Put* $N = N^A := \min\{f(x) \mid x \in \mathbb{Z},\ x \geq \max\{0, [M] + 1\}\}$. *Then we have*

$$N = \begin{cases} f([M] + 1) & \text{if } M > 0, \\ f(0) & \text{if } M < 0. \end{cases}$$

Proof For $x(>0) \in \mathbb{R}$, define

$$g(x) := 7\log x + \frac{8}{x}.$$

Since

$$g'(x) = \frac{7}{x} - \frac{8}{x^2} = \frac{7x-8}{x^2},$$

$g(x)$ is strictly monotonously increasing in $(8/7, \infty)$. Then for any $x(\geq 0) \in \mathbb{R}$, we have

$$Ix + J + 1 \geq J + 1 = b^A + \varepsilon^A + 1 \geq 3 > \frac{8}{7},$$

and so,

$$g(Ix + J + 1) \geq g(3) = 7\log 3 + \frac{8}{3} = 10.3569\cdots > 8 > 8 - \frac{7L}{2}.$$

Then we have

$$\begin{aligned} f'(x) &= \frac{(7/8)(Ix+J)^{-1/8} I\{2\log(Ix+J+1)+L\} - (Ix+J)^{7/8}\frac{2I}{Ix+J+1}}{\{2\log(Ix+J+1)+L\}^2} \\ &= \frac{I\left\{14\log(Ix+J+1)+7L-16\frac{Ix+J}{Ix+J+1}\right\}}{8(Ix+J)^{1/8}\{2\log(Ix+J+1)+L\}^2} \\ &= \frac{I\left\{14\log(Ix+J+1)+\frac{16}{Ix+J+1}+7L-16\right\}}{8(Ix+J)^{1/8}\{2\log(Ix+J+1)+L\}^2} \\ &= \frac{I(2g(Ix+J+1)+7L-16)}{8(Ix+J)^{1/8}\{2\log(Ix+J+1)+L\}^2} \\ &> 0, \end{aligned}$$

and hence $f(x)$ is strictly monotonously increasing in $[0, \infty)$.

Therefore, if $M > 0$, then $[M] + 1 \geq 1$, and so,

$$N = f([M] + 1).$$

If $M < 0$, then $[M] + 1 \leq 0$, and so,

$$N = f(0).$$

The proof is completed. □

Proof of Theorem 2 Let t be a positive integer with $t \geq \max\{1, [M] + 2\}$ and assume that $d := d(t)$ is square free. Put $x := t - 1$. Then we have $x \geq \max\{0, [M] + 1\}$ and $x \in \mathbb{Z}$. It follows from the definitions of I and J that

$$b + \varepsilon^A = b(t) + \varepsilon^A = q_{m+1}(t-1) + b^A + \varepsilon^A = Ix + J.$$

From this, together with (2.3) and $x \geq [M] + 1 > M = (e^8 - J)/I$, we have

$$d > (b + \varepsilon^A)^2 = (Ix + J)^2 > e^{16}. \tag{2.6}$$

On the other hand, we see from the definition of L and Lemma 2.1 that

$$\log \varepsilon_d < 2\log(Ix + J + 1) + L. \tag{2.7}$$

For $d \in \{d(t) \mid t \in \mathbb{Z},\ t \geq \max\{1, [M]+2\}, d(t) : \text{square-free}\}$, therefore, it follows from (2.1), (2.6), and (2.7) that

$$h_d > \frac{0.3275}{16} \cdot \frac{d^{7/16}}{\log \varepsilon_d} > \frac{0.3275}{16} \cdot \frac{(Ix + J)^{7/8}}{2\log(Ix + J + 1) + L} = \frac{0.3275}{16} f(x) \geq \frac{0.3275}{16} N$$

with one more possible exception. Hence Lemma 2.2 reads Theorem 2. □

Set

$$a := \max\{a_1, a_2, \ldots, a_m\}.$$

Let us give an upper bound for L.

Lemma 2.3 *Under the above setting, the following holds:*

$$L < 2(m-1)\log(a^2 + 1) - \sum_{k=2}^{m-1} 2\log a_k + 2\log 2.$$

Proof For simplify, set

$$B := \prod_{k=1}^{m} a_k \cdot \prod_{k=2}^{m} \left(1 + \frac{1}{a_{k-1}a_k}\right).$$

Then $L = \log(2B^2 + 1)$. For $x (>0) \in \mathbb{R}$, define $\varphi(x) := \log(x+1) - \log x$. Since

$$\varphi'(x) = \frac{1}{x+1} - \frac{1}{x} = -\frac{1}{x(x+1)} < 0,$$

$\varphi(x)$ is strictly monotonously decreasing in $(0, \infty)$. Therefore, when $x > 1$, we have

$$\varphi(x) < \varphi(1) = \log 2.$$

Thus

$$\log(x+1) < \log x + \log 2 \quad (x > 1). \tag{2.8}$$

Since $a_{k-1}a_k \leq a^2$, we obtain

$$\begin{aligned}\log B &= \sum_{k=1}^{m} \log a_k + \sum_{k=2}^{m} \log\left(1 + \frac{1}{a_{k-1}a_k}\right) \\ &= \sum_{k=1}^{m} \log a_k + \sum_{k=2}^{m} \log\left(\frac{a_{k-1}a_k + 1}{a_{k-1}a_k}\right) \\ &\leq \sum_{k=1}^{m} \log a_k - \sum_{k=2}^{m} \log a_{k-1} - \sum_{k=2}^{m} \log a_k + \sum_{k=2}^{m} \log(a^2 + 1) \\ &= \log a_m - \sum_{k=2}^{m} \log a_k + (m-1)\log(a^2+1) \\ &= (m-1)\log(a^2+1) - \sum_{k=2}^{m-1} \log a_k.\end{aligned}$$

Then by (2.8), we get

$$\begin{aligned}L = \log(2B^2+1) &< \log(2B^2) + \log 2 = 2\log B + 2\log 2 \\ &\leq 2(m-1)\log(a^2+1) - \sum_{k=2}^{m-1} 2\log a_k + 2\log 2,\end{aligned}$$

as desired. □

3 Proof of Theorem 1

In this section, we will prove Theorem 1.

Proposition 3.1 *Let $A = \langle a_1, \ldots, a_m \rangle$ be of pre-ELE type with length $m(\geq 4)$ such that $a_2 = a_3 = \cdots = a_{m-1} = 1$. Then $A = \langle 3, 1, 1, 1 \rangle$ or $\langle 4, 1, 1, 1, 1 \rangle$.*

Proof The proof is omitted. It needs the properties of "growth transformations" introduced in [2]. □

Lemma 3.2 *Let $A = \langle a_1, \ldots, a_m \rangle$ be of pre-ELE type with length $m(\geq 4)$. If $A \neq \langle 3, 1, 1, 1 \rangle, \langle 4, 1, 1, 1, 1 \rangle$, then*

$$\sum_{k=2}^{m-1} \log a_k \geq \log 2. \tag{3.1}$$

If $A = \langle 3, 1, 1, 1 \rangle$ (resp. $A = \langle 4, 1, 1, 1, 1 \rangle$), then

$$J = 9 < e^8 \ (resp.\ J = 16 < e^8). \tag{3.2}$$

Proof By Proposition 3.1, if $A \neq \langle 3, 1, 1, 1\rangle, \langle 4, 1, 1, 1, 1\rangle$, then (3.1) holds.
Let $A = \langle 3, 1, 1, 1\rangle$. Then the following holds:

n	0	1	2	3	4	5
q_n	0	1	3	4	7	11
r_n	1	0	1	1	2	3

Hence we have $\varepsilon^A = 1$ and $b^A = 8$, and so (3.2) holds.
Let $A = \langle 4, 1, 1, 1, 1\rangle$. Then the following holds:

n	0	1	2	3	4	5	6
q_n	0	1	4	5	9	14	23
r_n	1	0	1	1	2	3	5

Hence we have $\varepsilon^A = 1$ and $b^A = 15$, and so (3.2) also holds. □

Proof of Theorem 1 Assume that (1.3) holds. Then we have

$$\frac{0.3275}{16} J^{7/8} - 2\log(J+1) \geq 2(m-1)\log(4J^2/9+1). \tag{3.3}$$

Now we put

$$a := \max\{a_1, a_2, \ldots, a_m\}.$$

Then it holds in general that $a < 2[\sqrt{d(1)}]/3$ (see, for example, Perron [5, Satz 3.14]). Since $[\sqrt{d(1)}] = b^A + \varepsilon^A$, we have

$$(0 <)\, a < \frac{2}{3}[\sqrt{d(1)}] = \frac{2}{3}(b^A + \varepsilon^A) = \frac{2}{3}J.$$

Since $\log(x^2+1)$ is strictly monotonously increasing in $[0, \infty)$, we see from Lemmas 3.2 and 2.3 that

$$\begin{aligned} 2(m-1)\log(4J^2/9+1) &= 2(m-1)\log((2J/3)^2+1) \\ &> 2(m-1)\log(a^2+1) \\ &\geq 2(m-1)\log(a^2+1) - \sum_{k=2}^{m-1} 2\log a_k + 2\log 2 \\ &> L. \end{aligned} \tag{3.4}$$

By (3.3) and (3.4), therefore, we get

$$\frac{0.3275}{16} J^{7/8} - 2\log(J+1) > L.$$

From this inequality and (2.5), we have

$$h_d > \frac{0.3275}{16} f(0) = \frac{0.3275}{16} \cdot \frac{J^{7/8}}{2\log(J+1)+L} > 1$$

with one more possible exception, under the assumption $J > e^8$. □

4 A Sequence $\langle 2, \ldots, 2, 2, 1\rangle$ of Pre-ELE$_2$ Type

In this section, we consider a sequence

$$A = \langle 2, \ldots, 2, 2, 1\rangle$$

of pre-ELE$_2$ type with length $m(\geq 3)$. Then we have $\varepsilon^A = 1$, $I = q_{m+1}$, $b^A = q_{m+1}$, $J = b^A + \varepsilon^A = q_{m+1} + 1$, $a = 2$ and

$$d(t) = q_{m+1}^2 t^2 + 4q_m t + 2$$

for positive integer $t \geq 1$ (cf. [2, Lemma 2.1 (2)]). Moreover, we have

$$q_n = \frac{(1+\sqrt{2})^n - (1-\sqrt{2})^n}{2\sqrt{2}} \quad (1 \leq n \leq m),$$

and so

$$\begin{aligned} q_{m+1} &= q_m + q_{m-1} \\ &= \frac{(1+\sqrt{2})^m - (1-\sqrt{2})^m}{2\sqrt{2}} + \frac{(1+\sqrt{2})^{m-1} - (1-\sqrt{2})^{m-1}}{2\sqrt{2}} \\ &= \frac{(1+\sqrt{2})^m}{2\sqrt{2}}\left(1 + \frac{1}{1+\sqrt{2}}\right) - \frac{(1-\sqrt{2})^m}{2\sqrt{2}}\left(1 + \frac{1}{1-\sqrt{2}}\right) \\ &= \frac{(1+\sqrt{2})^m}{2} + \frac{(1-\sqrt{2})^m}{2}. \end{aligned} \tag{4.1}$$

For $m(\geq 3) \in \mathbb{Z}$, define

$$\theta(m) := 2(m-1)\log 5 - 2(m-3)\log 2.$$

Then by Lemma 2.3, we have $L < \theta(m)$. By putting

$$B_m := \frac{0.3275}{16} \cdot \frac{J^{7/8}}{2\log(J+1)+\theta(m)},$$

therefore, we obtain

$$\frac{0.3275}{16} f(0) = \frac{0.3275}{16} \cdot \frac{J^{7/8}}{2\log(J+1)+L} > B_m. \tag{4.2}$$

Example 4.1 Let $m = 11$. Then we have

$$J = J^A = 8120 > e^8,$$
$$\theta(11) = 20\log 5 - 16\log 2,$$

and hence,

$$B_{11} = \frac{0.3275}{16} \cdot \frac{8120^{7/8}}{2\log(8121)+20\log 5 - 16\log 2} = 1.3795\cdots.$$

Thus by Theorem 2 and (4.2), for $d \in \{d(t) \mid t \in \mathbb{Z},\ t \geq 1, d(t) : \text{square-free}\}$, we have

$$h_d > \frac{0.3275}{16} f(0) > B_{11} = 1.3795\cdots > 1$$

with one more possible exception.

By a similar calculation, we obtain the following:

m	10	12	13	14
J^A	3364	19602	47322	114244
B_m	$0.7026\cdots$	$2.7317\cdots$	$5.4482\cdots$	$10.9319\cdots$

Thus, as for $11 \leq m \leq 14$, we get a nontrivial lower bound for $h_{d(t)}$.

Now we will investigate when both (1.3) and $J > e^8$ hold. Note that

$$(1.3) \iff CJ^{7/8} - \log(J+1) - (m-1)\log(4J^2/9+1) > 0.$$

Since

$$\left|\frac{(1-\sqrt{2})^m}{2}\right| < 1,$$

it follows from (4.1) that

$$\frac{(1+\sqrt{2})^m}{2} - 1 < q_{m+1} < \frac{(1+\sqrt{2})^m}{2} + 1.$$

Then by putting

$$K := \frac{(1+\sqrt{2})^m}{2} + 1,$$

we have

$$K - 1 - 1 < q_{m+1} < K - 1 + 1,$$

that is,

$$K - 1 < J < K + 1.$$

Hence we have

$$\begin{aligned}
&CJ^{7/8} - \log(J+1) - (m-1)\log(4J^2/9+1) \\
&\quad > C(K-1)^{7/8} - \log(K+2) - (m-1)\log(4(K+1)^2/9+1) \\
&\quad = C(K-1)^{7/8} - \log(K+2) - m\log(4(K+1)^2/9+1) + \log(4(K+1)^2/9+1) \\
&\quad = C(K-1)^{7/8} - m\log\left(\frac{4}{9}K^2 + \frac{8}{9}K + \frac{13}{9}\right) + \log\left(\frac{\frac{4}{9}K^2 + \frac{8}{9}K + \frac{13}{9}}{K+2}\right) \\
&\quad = C(K-1)^{7/8} - m\log\left(\frac{4}{9}K^2 + \frac{8}{9}K + \frac{13}{9}\right) + \log\left(\frac{4}{9}(K-1) + \frac{4K+21}{9(K+2)}\right).
\end{aligned}$$

Here, if $m \geq 5$, then $K \geq 35$, and so

$$\frac{1}{2}(K-1)^2 - \left(\frac{4}{9}K^2 + \frac{8}{9}K + \frac{13}{9}\right) = \frac{1}{18}\{(K-17)^2 - 306\} > 0.$$

Then we get

$$\begin{aligned}
&CJ^{7/8} - \log(J+1) - (m-1)\log(4J^2/9+1) \\
&\quad > C(K-1)^{7/8} - m\log\left(\frac{1}{2}(K-1)^2\right) + \log\left(\frac{4}{9}(K-1)\right) \\
&\quad = C\left(\frac{(1+\sqrt{2})^m}{2}\right)^{7/8} - m\log\left(\frac{(1+\sqrt{2})^{2m}}{8}\right) + \log\left(\frac{2(1+\sqrt{2})^m}{9}\right) \\
&\quad = \left(\frac{1}{2}\right)^{7/8} C(1+\sqrt{2})^{7m/8} - 2m^2\log(1+\sqrt{2}) \\
&\qquad + m\log 8 + m\log(1+\sqrt{2}) + \log\left(\frac{2}{9}\right).
\end{aligned}$$

For $x(\geq 0) \in \mathbb{R}$, define

$$\psi(x) := \left(\frac{1}{2}\right)^{7/8} C(1+\sqrt{2})^{7x/8} - 2x^2 \log(1+\sqrt{2}) + x \log 8 + x \log(1+\sqrt{2}) + \log\left(\frac{2}{9}\right).$$

Then we have

$$\psi'(x) = \frac{7}{8}\left(\frac{1}{2}\right)^{7/8} C(1+\sqrt{2})^{7x/8} \log(1+\sqrt{2}) - 4x \log(1+\sqrt{2}) + \log 8 + \log(1+\sqrt{2}),$$

$$\psi''(x) = \left(\frac{7}{8}\right)^2 \left(\frac{1}{2}\right)^{7/8} C(1+\sqrt{2})^{7x/8} \log(1+\sqrt{2})^2 - 4\log(1+\sqrt{2})$$
$$= \log(1+\sqrt{2}) \left\{ \left(\frac{7}{8}\right)^2 \left(\frac{1}{2}\right)^{7/8} C(1+\sqrt{2})^{7x/8} \log(1+\sqrt{2}) - 4 \right\}.$$

Since

$$\psi'(11) = -15.0205\cdots < 0,$$
$$\psi'(12) = 5.6300\cdots > 0$$

and $\psi''(x) > 0$ if $x \geq 10$, $\psi(x)$ is strictly monotonously increasing in $[12, \infty)$. Then by

$$\psi(14) = -32.8654\cdots < 0,$$
$$\psi(15) = 235.9345\cdots > 0,$$

we have $\psi(x) > 0$ if $x \geq 15$. Namely, for any integer $m \geq 15$, we have

$$CJ^{7/8} - \log(J+1) - (m-1)\log(4J^2/9+1) \geq 0.$$

On the other hand, it follows from $J = q_{m+1} + 1$ and $e^8 = 2980.9579\cdots$ that $J > e^8$ holds when $m \geq 10$. Then combine with Example 4.1 we get the following.

Proposition 4.1 *Let $m \geq 11$ and let $A = \langle 2, \ldots, 2, 2, 1\rangle$ with length m. Then for $d \in \{d(t) \mid t \in \mathbb{Z},\ t \geq 1, d(t) : \text{square-free}\}$, we have $h_d > 1$ with one more possible exception.*

References

1. F. Kawamoto, Y. Kishi, K. Tomita, Continued fraction expansions with even period and primary symmetric parts with extremely large end. Comm. Math. Univ. Sancti Pauli **64**(2), 131–155 (2015)
2. F. Kawamoto, Y. Kishi, H. Suzuki, K. Tomita, Real quadratic fields, continued fractions, and a construction of primary symmetric parts of ELE type. Kyushu J. Math. **73**(1), 165–187 (2019)
3. F. Kawamoto, K. Tomita, Continued fractions and certain real quadratic fields of minimal type. J. Math. Soc. Japan **60**(3), 865–903 (2008)
4. S. Lang, *Introduction to Diophantine Approximations* (Addison-Wesley, Reading, MA, 1966)
5. O. Perron, Die Lehre von den Kettenbrüechen, Band I: Elementare Kettenbrüche, 3te Aufl. B. G. Teubner Verlagsgesellschaft, Stuttgart (1954)
6. T. Tatuzawa, On a theorem of Siegel. Jpn. J. Math. **21**, 163–178 (1951)
7. H. Yokoi, New invariants and class number problem in real quadratic fields. Nagoya Math. J. **132**, 175–197 (1993)

A Survey of Certain Euclidean Number Fields

Kotyada Srinivas and Muthukrishnan Subramani

2010 Mathematics Subject Classification 11A05 (primary) · 11R04 (secondary).

1 Introduction

Let K be a number field and $\mathcal{O}_K$ be its ring of integers. We say that K (or $\mathcal{O}_K$) is Euclidean if there exists a (Euclidean) function $\phi : \mathcal{O}_K \rightarrow \mathbb{N} \cup \{0\}$ satisfying the conditions (i) $\phi(\alpha) = 0$ if and only if $\alpha = 0$, and (ii) for all $\alpha, \beta \neq 0 \in \mathcal{O}_K$ there exists a $\gamma \in \mathcal{O}_K$ such that $\phi(\alpha - \beta\gamma) < \phi(\beta)$. For example the ring of rational integers $\mathbb{Z}$ is Euclidean with respect to the absolute value function $\phi(a) = |a|$. As an immediate consequence, we obtain the fundamental theorem of arithmetic. Also, if an integral domain admits a Euclidean function, i.e., if it a Euclidean domain, then it is a principal ideal domain. Therefore, a natural question arises: *Given a number field K (or an integral domain D) what is the criteria to decide whether it is Euclidean or not? Is there a classification of all Euclidean number fields?*

In this chapter we shall address such questions and give an update of recent results in this direction.

This note is a talk given in the conference ICCGNFRT meeting at HRI, Allahabad in September 2017.

K. Srinivas (✉)
Institute of Mathematical Sciences, HBNI, CIT Campus Taramani,
Chennai 600 113, India
e-mail: srini@imsc.res.in

M. Subramani
Harishchandra Research Institute, HBNI, Chhatnag Road,
Allahabad 211 019, India
e-mail: msubramani@hri.ac.in

K. Chakraborty et al. (eds.), *Class Groups of Number Fields and Related Topics*,
https://doi.org/10.1007/978-981-15-1514-9_5

To begin with, let us first define the norm-Euclidean function for quadratic fields. Let m be a square free integer. Let $K = \mathbb{Q}(\sqrt{m})$. The integral domain $\mathcal{O}_K$ is called norm-Euclidean, if there is a function $\phi = \phi_m : \mathbb{Q}(\sqrt{m}) \to \mathbb{Q}$ defined by

$$\phi_m(r + s\sqrt{m}) = |r^2 - ms^2| \tag{1.1}$$

for all $r, s \in \mathbb{Q}$.

An elegant criteria to check the Euclidean property was given by Lenstra in [1] which we state below.

Proposition 1.1 *Let K be a quadratic field and $\mathcal{O}_K$ be its ring of integers. The norm map $N : \mathcal{O}_K \to \mathbb{N} \cup \{0\}$ is Euclidean if and only if for every $\alpha \in \mathcal{O}_K$ there exists $a, \beta \in K$ such that $N(\alpha - \beta) < 1$.*

The proposition 1.1 is powerful to classify all norm-Euclidean imaginary fields, since the norm on imaginary quadratic fields assumes only positive integer values. Consequently, all norm Euclidean quadratics fields were classified, which are as follows (see also [2]).

If $m < 0$, then $\mathbb{Z} + \mathbb{Z}\sqrt{m}$ is norm-Euclidean if and only if $m = -1, -2$ and if $m \equiv 1 \pmod 4$, then the integral domain $\mathbb{Z} + \mathbb{Z}(1 + \sqrt{m})/2$ is norm-Euclidean if and only if $m = -3, -7, -11$.

The story for real case is a bit tricky. This is mainly due to the fact that norm assumes both negative and positive integer values. But the following reinterpretation of the definition turns out to be useful.

Definition 1.1 For $\alpha \in K$, let $m(\alpha) = \inf\{N(\alpha - \beta) : \beta \in \mathcal{O}_K)\}$, it is called the Euclidean minimum of K at α. The Euclidean minimum of K is defined as $M(K) := \sup\{m(\alpha) : \alpha \in K\}$.

It follows from the Proposition 1.1 that $M(K) < 1$ if and only if K is norm-Euclidean. A bound for $M(K)$ is given by the following theorem of Cassels.

Theorem 1.1 (Cassels [3]) *Let $K = \mathbb{Q}(\sqrt{m})$ be a real quadratic field with discriminant d, then*

$$\frac{\sqrt{m}}{16 + 6\sqrt{6}} \le M(K) \le \frac{\sqrt{m}}{4} \tag{1.2}$$

Observe that if the discriminant m is sufficiently large then $M(K) > 1$ and thus K will not be norm-Euclidean. From this it follows that the number of norm-Euclidean real quadratic fields is finite. On the other hand to determine the positive squre-free integers m for which the integral domains $\mathbb{Z} + \mathbb{Z}\sqrt{m}$ ($m \equiv 2, 3 \pmod 4$) and $\mathbb{Z} + \mathbb{Z}(1 + \sqrt{m})/2$ ($m \equiv 1 \pmod 4$) are norm-Euclidean took considerable efforts of numerous mathematicians, including, Dickson, Perron, Oppenheim, Erdös, Davenport, Chatland. It must be pointed out that using only the criteria in the Proposition 1.1, one can show that certain real quadratic fields are not norm-Euclidean, for example $\mathbb{Q}(\sqrt{6})$. In this direction, the following Lemma 1.1 turns out to be useful in the search of Euclidean real quadratic fields.

Lemma 1.1 *Let K be a quadratic field of discriminant D and h_K be its class number. Suppose D has r distinct prime factors, then $2^{r-1}|h_K$ if $D<0$ and $2^{r-2}|h_K$ if $D>0$.*

The above lemma helps in restricting our attention to only certain quadratic fields. In [4], Erdös and Ko proved that the quadratic fields of the form $\mathbb{Q}(\sqrt{p})$ are not norm-Euclidean for all sufficiently large primes p. Huo Loo-Keng [5] made it more precise.

Theorem 1.2 (Huo [5]) *The quadratic field $\mathbb{Q}(\sqrt{p})$ is not norm-Euclidean if $p > e^{250}$.*

By reducing this upper bound, the problem of classifying all norm-Euclidean real quadratic fields was finally completed by Chatland and Davenport (see [6]). Their result is as follows:

If m is a positive squarefree integer with $m \equiv 2, 3 \pmod 4$, then $\mathbb{Z}+\mathbb{Z}\sqrt{m}$ is norm-Euclidean if and only if $m = 2, 3, 6, 7, 11, 19$ and if $m \equiv 1 \pmod 4$, then the integral domain $\mathbb{Z}+\mathbb{Z}(1+\sqrt{m})/2$ is norm-Euclidean if and only if $m = 5, 13, 17, 21, 29, 33, 37, 41, 57, 73$.

This gives a complete classification of all norm-Euclidean reat quadratic fields.

Now, one can ask the following question: *Suppose K is known to be not norm Euclidean. Could it be Euclidean with respect to some other function ϕ?*

The answer to the above question is given in terms of existence of *universal side divisors*, which we define below.

Definition 1.2 Let D be an integral domain, $U(D)$ be the group of units in D. Set $\tilde{D} = U(D) \cup \{0\}$. An element $u \in D - \tilde{D}$ is called a universal side divisor if for any $x \in D$ there exists some $z \in \tilde{D}$ such that $u|x-z$.

Using this notion, Motzkin showed that if D is an integral domain that is not a field and if D has no side divisors then D is not Euclidean.

Using this criteria, one can show if $m \equiv 2, 3 \pmod 4$ and $m < -2$, then $\mathbb{Z}+\mathbb{Z}\sqrt{m}$ is not Euclidean. Similarly, if $m \equiv 1 \pmod 4$ and $m < -11$, then $\mathbb{Z}+\mathbb{Z}(1+\sqrt{m})/2$ is not Euclidean. Thus for $K = \mathbb{Q}(\sqrt{m})$, m squarefree negative integer, the integral domain $\mathcal{O}_K$ is Euclidean if and only if $m = -1, -2, -3, -7, -11$.

However, the picture is not that rosy when $m > 0$! In fact, very little is known in this case. By means of explicit construction Clark [7] showed that $\mathbb{Z}+\mathbb{Z}((1+\sqrt{69})/2)$ is Euclidean with respect to the function

$$\phi\left(a+b\left(\frac{1+\sqrt{69}}{2}\right)\right) = \begin{cases} |a^2+ab-17b^2|, & \text{if } (a,b) \neq (10,3), \\ 26, & \text{if } (a,b) = (10,3), \end{cases}$$

though it is known to be not norm-Euclidean. The first breakthrough came in 1949 through the work of Motzkin [8] who gave a beautiful criteria for a domain to be Euclidean.

Motzkin's criteria *Given a domain R, let $A_0 := \{0\} \cup R^{\times}$. Let*

$$A_i = A_{i-1} \cup \{a \in R | \forall x \in R, \exists y \in A_{i-1} : x - y \in (a)\}$$

for $i \geq 1$ and let

$$A := \cup_{i=0}^{\infty} A_i$$

If $A = R$, then R is Euclidean. In particular, the function ϕ_R defined by $\phi_R(\alpha) = 0$ if and only if $\alpha = 0$, and $\phi_R(\alpha) = i$ if $\alpha \in A_i - A_{i-1}$ is an Euclidean function.

It is to be remarked that Motzkin's construction is ubiquitous in all the subsequent progress in this area of research. This idea enabled Weinberger [9] in 1973, in conjunction with Generalized Riemann Hypothesis (GRH), to show that all algebraic number fields with infinitely many units and whose ring of integers are PIDs are in fact Euclidean! In 1977 Lenstra [1] extended this result to the ring of S integers in a number field under GRH. We define the notion of S integers for the sake of completion.

Definition 1.3 Let S be a finite set of places of K containing the infinite places S_{∞}. An element x of K is called an S-integer if $\text{ord}_{\mathfrak{p}}(x) \geq 0$ for all primes $\mathfrak{p}$ of K not in S. The ring of S integers will be denoted by $\mathcal{O}_S$.

In 1985 Gupta, Ram Murty, Kumar Murty managed to remove GRH from Lenstra's result. Their result reads as follows.

Theorem 1.3 (Gupta, Murty, Murty [10]) *Let K be Galois over $\mathbb{Q}$ such that*

(1) $|S| \geq max\{5, 2[K : \mathbb{Q}] - 3\}$;
(2) K has a real embedding or $\zeta_g \in K$.

If $\mathcal{O}_S$ is a PID, then it is Euclidean. Here $g := \gcd\{N_{K/\mathbb{Q}}(\mathfrak{p}) - 1 : \mathfrak{p} \in S - S_{\infty}\}$.

In 1995, Ram Murty and Clark [11] showed how to remove GRH for large class of algebraic number fields by introducing the concept of *admissible* primes, which is an indispensable ingredient in the determination of Euclidean algorithm for abelian algebraic number fields. We now introduce the concept of *admissible* primes.

Definition 1.4 Let K be an algebraic number field with class number one. Suppose that $\pi_1, \pi_2, \ldots, \pi_t$ are distinct, unramified primes with inertial degree one lying above odd rational primes. Then the set $\{\pi_1, \pi_2, \ldots, \pi_t\}$ is called an admissible set of primes if, for all $\beta = \pi_1^{a_1} \pi_2^{a_2} \ldots \pi_t^{a_t}$ with $a_i \in \mathbb{N} \cup \{0\}$, every element in the coprime residue class $\pmod{\beta}$ is represented by a unit in $\mathcal{O}_K^{\times}$. i.e., the natural canonical map $\mathcal{O}_K^{\times} \to (\mathcal{O}_K/\beta)^{\times}$ is surjective.

In the same paper, they also established that it is enough to take all $a_i's$ to be equal to 2. Their theorem runs as follows.

Theorem 1.4 (Ram Murty and Clark [11]) *Let K be a totally real Galois extension with degree n_K such that $\mathcal{O}_K$ has class number one. Suppose that $\mathcal{O}_K$ has a set S of admissible primes with $m = |n_K - 4| + 1$ elements, then $\mathcal{O}_K$ is Euclidean.*

Remark This theorem is handy when the rank of the unit group is large. However, when K is a real quadratic field or cyclic cubic field, the theorem demands the existence of an admissible set with 3 elements. This is not possible for real quadratic fields, as can be seen by the following arguments.

By Dirichlet unit theorem, the group of units, $\mathcal{O}_K^\times \cong U_K \times \mathbb{Z}^r$, where U_K is the torsion group and r is the rank of K. Since the unit group $\mathcal{O}_K^\times$ is generated by $r+1$ elements, the multiplicative group $(\mathcal{O}_K/\pi_1^{a_1}\pi_2^{a_2}\dots\pi_t^{a_t})^\times$ is also generated by $r+1$ elements.

On the other hand, by Chinese remainder theorem, we have

$$(\mathcal{O}_K/\pi_1^{a_1}\pi_2^{a_2}\dots\pi_t^{a_t})^\times \cong \oplus_{i=1}^t(\mathcal{O}_K/\pi_i^{a_i})^\times. \tag{1.3}$$

This implies $t \le r+1$. This says that when K is real quadratic (unit rank 1), we can have at the most 2 elements in the admissible set and when K is cyclic cubic (unit rank 2), the admissible set may contain at the most 3 elements. Thus for real quadratic case, the above theorem is not useful!

In order to deal with number fields with small degrees Ram Murty and Harper [12], considered a variation of Motzkin's construction which is as follows.

Variation of Motzkin's construction: Let B_0 be the monoid generated by the unit group and an admissible set of primes. For $n \ge 1$, define the sets B_n as follows:

$$B_n := \{\pi - \text{prime} \in \mathcal{O}_K^\times : B_0 \cup B_{n-1} \to (\mathcal{O}_K/\pi)^\times \text{is surjective}\}$$

and let $B := \cup_{n=0}^\infty B_n$. If all primes of $\mathcal{O}_K$ lies in B and K has class number one, then K is Euclidean.

These modified criteria allowed them to prove the following important results.

Theorem 1.5 (Harper and Murty [12]) *Suppose that $\mathcal{O}_K$ is a PID. Let $\mathbb{B}_1(x)$ denote the cardinality of the set of all elements in B_n whose norm is less than or equal to x. If*

$$\mathbb{B}_1(x) \gg \frac{x}{log^2 x}$$

then $\mathcal{O}_K$ is Euclidean.

Theorem 1.6 (Harper and Murty [12]) *If $K/\mathbb{Q}$ is a finite Galois extension with unit rank > 3, then $\mathcal{O}_K$ is Euclidean if and only if $\mathcal{O}_K$ is a PID.*

Theorem 1.7 (Harper and Murty [12]) *Let $K/\mathbb{Q}$ be abelian of degree n with $\mathcal{O}_K$ having class number one, that contains a set of admissible primes with s elements. Let r be the rank of the unit group. If $r+s \ge 3$, then $\mathcal{O}_K$ is Euclidean.*

Remark For real quadratic fields, Theorem 1.7 says that it is enough to find an admissible set with two elements. Indeed, Harper in 2004 [13] showed that $\mathbb{Z}[\sqrt{14}]$ is Euclidean by exhibiting the admissible set $\{5-\sqrt{14}, 3-2\sqrt{14}\}$. This was a great success since it was known before that $\mathbb{Q}(\sqrt{14})$ is not norm Euclidean and it was the field with smallest discriminant whose Euclidean nature was evading detection! He further proved in the same paper that all real quadratic fields with discriminant ≤ 500 and having class number one are Euclidean.

Without the strong assumption of GRH, under some reasonably plausible (in terms of heuristics) Hardy-Littlewood and Wieferich primes conjecture, the authors and Ram Murty in [14] proved that if K is real quadratic, then it has an admissible set with two elements. We would like to state the main results of [14].

Conjecture 1.2 (Hardy-Littlewood conjecture) *Fix a natural number r and b coprime to r. Hardy and Littlewood conjectured that the number of primes $p \leq x$ with $p \equiv b \pmod r$ such that $2p + 1$ is also prime is*

$$\gg \frac{x}{\log^2 x}.$$

Conjecture 1.3 (Wieferich primes conjecture) *Let ϵ be an element of $\mathcal{O}_K^\times$ of infinite order. The number of primes $p \leq x$ such that*

$$\epsilon^{p-1} \equiv 1 \pmod{p^2}$$

is $o(x/\log^2 x)$.

The main result is

Theorem 1.8 (Ram Murty, Srinivas, Subramani [14]) *Assume the Hardy-Littlewood and the Wieferich primes conjectures. If K is a real quadratic field such that $\mathcal{O}_K$ has class number one, then $\mathcal{O}_K$ is Euclidean.*

Remark The key result used in establishing the main Theorem 1.8 is the following lemma.

Lemma 1.2 *Let L be a number field, $\mathcal{O}_L$ be its ring of integers and let $\epsilon \in \mathcal{O}_L$ be a unit of infinite order. If $\mathfrak{q}_1$ and $\mathfrak{q}_2$ are distinct, unramified prime ideals with odd prime norms q_1 and q_2 and if*

(1) ϵ has order $q_1(q_1 - 1)/2$ modulo $\mathfrak{q}_1^2$;
(2) $q_1 \equiv 3 \pmod 4$;
(3) $\gcd(q_1(q_1 - 1)/2, q_2(q_2 - 1)) = 1$; and
(4) ϵ has the order $q_2(q_2 - 1)$ modulo $\mathfrak{q}_2^2$;

then $\mathcal{O}_L^\times$ maps onto $(\mathcal{O}_L/\mathfrak{q}_1^2\mathfrak{q}_2^2)^\times$.

Remark The Conjectures 1.2 and 1.3 allows us to choose primes q_1 and q_2 satisfying all the conditions of Lemma 1.2, thereby yielding an admissible set $\{\mathfrak{q}_1, \mathfrak{q}_2\}$. Now invoking Theorem 1.7 established the result. For details see [14].

2 Explicit Construction of Potentially Euclidean Real Quadratic Fiel

As mentioned in the remark following Theorem 1.7, Harper [13] showed that all real quadratic fields with discriminant ≤500 and having class number one are Euclidean. This he did by producing an admissible set with two elements for each of the fields. Motivated by this idea, the authors of this paper and Ram Murty [14] constructed an infinite family of real quadratic fields K such that the class number of K is one if and only if $\mathcal{O}_K$ is Euclidean. We briefly describe the construction below, for details see [14].

Fix two primes $p_1 := 11$ and $p_2 := 13$. Let

$$d := (a+1)^2 b^2 n^2 + 2(a+1)^2 n + 23 \tag{2.1}$$

where a, b, n are integers such that

$$a \equiv 24 \pmod{p_1^3 p_2^3},\ b \equiv 5 \pmod{p_1^3 p_2^3}, \tag{2.2}$$

and

$$n \equiv 0 \pmod{p_1 p_2}. \tag{2.3}$$

We define

$$K := \mathbb{Q}(\sqrt{d}) = \mathbb{Q}\left(\sqrt{(a+1)^2 b^2 n^2 + 2(a+1)^2 n + 23}\right). \tag{2.4}$$

With the above notations, the main results are as follows.

Theorem 2.1 *Let $K = \mathbb{Q}(\sqrt{d})$ be as defined above. Then there exists a set $\{\mathfrak{p}_1, \mathfrak{p}_2\}$ of two unramified prime ideals with odd prime norms p_1 and p_2 respectively such that the canonical map $\mathcal{O}_K^\times \to (\mathcal{O}_K/\mathfrak{p}_1^2\mathfrak{p}_2^2)^\times$ is surjective.*

As a consequence of Theorem 2.1, we deduce the following:

Theorem 2.2 *There exists a family $C := \{\mathbb{Q}(\sqrt{d}) : d \text{ is prime}\}$ of real quadratic fields such that $\mathcal{O}_K$ is Euclidean if and only if it has class number one.*

Remark The primes $p_1 = 11$ and $p_2 = 13$ were chosen to illustrate how to produce the admissible set. In fact, the main theorem holds true for any real quadratic field K provided there exists two unramified rational primes $p_1 \equiv 3 \pmod 4$ and $p_2 \equiv 1 \pmod 4$ satisfying the following conditions:

(1) $N(\epsilon^{p_1(p_1-1)}) \not\equiv 0 \pmod{p_1}$;
(2) $N(\epsilon^{p_2(p_2-1)}) \not\equiv 0 \pmod{p_2}$; and
(3) $gcd(p_1(p_1-1)/2, p_2(p_2-1)) = 1$;

for any fixed unit $\epsilon \in \mathcal{O}_K^\times$.

3 The Cubic Case

Now we shall briefly mention the results known in the case of cyclic cubic number fields. Let K be a cyclic cubic field with discriminant f^2, where f is the conductor of K. The study of norm-Euclidean cubic fields was initiated by Heilbronn. In [15], he proved that only finitely many such cyclic cubic fields are Euclidean. In 1969, Smith [16] proved that the cyclic cubic fields with conductors 7, 9, …, 67 are norm-Euclidean. Further, in the same paper he showed that the fields with conductors 73, 79, 97, 139, 151 and $163 < f < 10^4$ are *not* norm-Euclidean. In [17], the authors proved the following theorem.

Theorem 3.1 *Let K be a cyclic cubic field with conductor f, satisfying $73 \leq f \leq 11971$ and let $\mathcal{O}_K$ be its ring of integers. Then $\mathcal{O}_K$ is Euclidean if and only if it has class number one.*

Remark K is cyclic cubic implies the Galois group of K over $\mathbb{Q}$ is cyclic of order three, therefore K is totally real. This, in turn means that the rank of unit group is 2. Thus, by Theorem 1.7 we need to exhibit an *admissible* set of primes with one element (i.e. $s = 1$). This is done explicitly for each conductor in the above range. A sage code is also available. For details see [17].

If K be a complex cubic field, an inequality of the type 1.2 exists for Euclidean minima, from which it follows that there are only finitely many complex cubic fields which are norm-Euclidean. We end this note by stating an important conjecture by Lemmermeyer [18].

Conjecture 3.1 *Let K be a complex cubic field and d be its discriminant. Then K is norm-Euclidean if and only $-d$ = 23, 31, 44, 59, 76, 83, 87, 104, 107, 108, 116, 135, 139, 140, 152, 172, 175, 200, 204, 211, 212, 216, 231, 239, 243, 244, 247, 255, 268, 300, 324, 356, 379, 411, 419, 424,431, 440, 451, 460, 472, 484, 492, 499, 503, 515, 516, 519, 543, 628, 652, 687, 696, 728, 744, 771, 815, 876.*

Acknowledgements The authors wish to thank the Organizers of ICCGNFRT–2017 for their kind invitation, for making the stay a happy and productive one.

References

1. H.W. Lenstra Jr., On Artin's conjecture and Euclid's algorithm. Inven. Math. **42**, 201–224 (1977)
2. S. Alaca, K.S. Williams, *Introductory Algebraic Number Theory* (Cambridge University Press, 2003)
3. J.W.S. Cassels, Yet another proof of Minkowskis theorem on the product of two inhomogeneous linear forms, in *Proceedings of the Cambridge Philosophical Society*, vol. 49 (1953), pp. 365–366
4. P. Erdös, Ch. Ko, Note on the Euclidean algorithm. J. London Math. Soc. **13**, 3–8 (1938)

5. H. Loo-Keng, On the distribution of quadratic nonresidues and the Euclidean algorithm in real quadratic fields. Trans. Amer. Math. Soc. **56**, 537–546 (1944)
6. H. Chatland, H. Davenport, Euclid's algorithm in quadratic number fields. Canad. J. Math. **2**, 289–296 (1950)
7. D.A. Clark, A quadratic field which is Euclidean but not norm-Euclidean. Manuscripta Mathematica **83**, 327–330 (1994)
8. T. Motzkin, The Euclidean algorithm, Bull. Am. Math. Soc. **55**(12), 1142–1146 (1949)
9. P.J. Weinberger, On Euclidean rings of algebraic integers, in *Proceedings of Symposia in Pure Mathematics, Analytic Number Theory, AMS*, vol. 24 (1973), pp. 321–332
10. R. Gupta, M. Ram Murty, V. Kumar Murty, The Euclidean algorithm for S-integers in number theory, in *CMS Conference on Proceedings of the American Mathematical Society* (*Montreal, June 1985*), vol. 7 (1987), pp. 189–201
11. D.A. Clark, M. Ram Murty, *The Euclidean algorithm for Galois extensions*. J. für die reine und angewandte Mathematik. **459**(1995), 151–162
12. M. Harper, M. Ram Murty Euclidean rings of algebraic integers. Canad. J. Math. **56**(1), 71–76 (2004)
13. M. Harper, $\mathbb{Z}[\sqrt{14}]$ is Euclidean. Canad. J. Math. **56**(1), 55–70 (2004)
14. M. Ram Murty, K. Srinivas, M. Subramani, Admissible primes and Euclidean quadratic fields, J. Ramanujan Math. Soc. **33**(2), 135–147 (2018)
15. H. Heilbronn, On Euclid's algorithm in cubic self-conjugate fields, in *Proceedings of the Cambridge Philosophical Society*, vol. 46 (1950), pp. 377–382
16. J.R. Smith, On Euclids algorithm in some cyclic cubic fields. J. London Math. Soc. **44**, 577–582 (1969)
17. K. Srinivas, M. Subramani, A note on Euclidean cyclic cubic fields. J. Ramanujan Math. Soc. **33**(2), 125–133 (2018)
18. F. Lemmermeyer, The Euclidean algorithm in algebraic number fields. Exposition. Math. **13**(5), 385–416 (1995)

Divisibility of Class Number of a Real Cubic or Quadratic Field and Its Fundamental Unit

Anupam Saikia

2010 Mathematics Subject Classification Primary 11R16 · Secondary 11R27 · 11R29 · 11R11

1 Introduction

First, we consider the fundamental unit of a pure cubic field $K = \mathbb{Q}(\sqrt[3]{m})$ where m is a square-free natural number not congruent to ± 1 modulo 9. Then it is well known that the ring of integers of K is $\mathcal{O}_K = \mathbb{Z}[\sqrt[3]{m}]$. Let h_m denote the class number of K and $\xi_m = x + y\sqrt[3]{m} + z\sqrt[3]{m^2}$ be the fundamental unit of K. We relate congruence properties of x, y and z to the class number of K. Consequently, one can obtain the following result.

Theorem 1.1 *If h_m is not divisible by 3, then m must be either p or $3p$ for some prime p.*

Theorem 1.1 is in agreement with an old result of Gerth ([2]). By Theorem 1.1, the class number of $\mathbb{Q}(\sqrt[3]{m})$ is divisible by 3 for any square-free composite number $m \equiv 2, 4, 5$ or $7 (mod\ 9)$.

Theorem 1.2 *Suppose $m = 3p$ where $p \neq 3$ is a prime. Let the fundamental unit of K be $\xi_m = x + y\sqrt[3]{m} + z\sqrt[3]{m^2}$. If 3 does not divide h_m then $x^2 \equiv 1\ mod\ 27p$ and $y \equiv z \equiv 0\ mod\ 3$.*

For $\mathbb{Q}(\sqrt[3]{p})$ too, we can similarly obtain Theorem 2.8 and more precisely, Propositions 2.6 and 2.7. Our approach is to exploit the ramified primes in $K/\mathbb{Q}$. We

A. Saikia (✉)
Department of Mathematics, Indian Institute of Technology,
Guwahati 781039, Assam, India
e-mail: a.saikia@iitg.ernet.in

K. Chakraborty et al. (eds.), *Class Groups of Number Fields and Related Topics*,
https://doi.org/10.1007/978-981-15-1514-9_6

later adopt the same approach to deduce congruence relations for the fundamental unit of a real quadratic field of odd class number (see Theorem 3.3). An immediate consequence of our approach is the classically well-known result that a real quadratic field with discriminant having more than or equal to three prime factors has even class number (see Corollary 3.2).

2 Fundamental Unit of $\mathbb{Q}(\sqrt[3]{m})$ when $3 \nmid h_m$

It is easily seen that the norm of ξ_m has to be positive, hence

$$\mathrm{Norm}_{K/\mathbb{Q}}(\xi_m) = x^3 + my^3 + m^2z^3 - 3mxyz = 1. \tag{2.1}$$

We can establish certain divisibility relations between ξ_m and ξ_m^2 as follows.

Lemma 2.1 *Suppose $K = \mathbb{Q}(\sqrt[3]{m})$ is a pure cubic field where m is not a multiple of 3. Let $\xi_m = x + y\sqrt[3]{m} + z\sqrt[3]{m^2} \in \mathbb{Z}[\sqrt[3]{m}]$ denote the fundamental unit of K and let $\xi_m^2 = x_1 + y_1\sqrt[3]{m} + z_1\sqrt[3]{m^2}$. Then 3 divides y, z if and only if 3 divides y_1, z_1.*

The above result can be explicitly obtained by simple computation as done in ([1]). One needs only to use (2.1) and the obvious relations

$$\begin{aligned} & x_1 + y_1\sqrt[3]{m} + z_1\sqrt[3]{m^2} = (x + y\sqrt[3]{m} + z\sqrt[3]{m^2})^2 \\ \implies \quad & x_1 = x^2 + 2myz, \quad y_1 = mz^2 + 2xy, \quad z_1 = y^2 + 2xz. \end{aligned} \tag{2.2}$$

The following lemma is crucial for the rest of the article.

Lemma 2.2 *Let $q \neq m$ be a prime that ramifies in $K = \mathbb{Q}(\sqrt[3]{m})$. If $3 \nmid h_m$ then either ξ_m or ξ_m^2 can be written as $\frac{\alpha^3}{q}$ for some $\alpha \in \mathcal{O}_K$.*

Proof As q ramifies in K, $q\mathcal{O}_K = \wp^3$ where $\wp$ is a prime ideal. Thus the ideal class of $\wp$ is an element of order dividing 3. Since the class number is not divisible by 3, $\wp$ must be principal. Hence $q\mathcal{O}_K = \wp^3 = \alpha^3\mathcal{O}_K$ for some $\alpha \in \mathcal{O}_K$. Therefore, $\pm\xi_m^j q = \alpha^3$. If $3 \mid j$, it would imply that $\sqrt[3]{q} \in K$ which leads to a contradiction as $q \neq m$. By modifying the element α suitably, we can take $j = 1$ or $j = 2$. □

Lemma 2.3 *If $3 \nmid h_m$ then m is either a prime or a multiple of 3.*

Proof Suppose m is neither a prime nor a multiple of 3. Let p be a (proper) prime factor of m. As p ramifies in $\mathbb{Q}(\sqrt[3]{m})$, $\xi_m^i = \frac{\beta^3}{p}$ by Lemma 2.2 where $i = 1$ or 2. By explicitly substituting $\beta = a_1 + b_1\sqrt[3]{m} + c_1\sqrt[3]{m^2}$, we find that y, z, y_1 and z_1 are all divisible by 3 by Lemma 2.1.

As 3 also ramifies in $\mathbb{Q}(\sqrt[3]{m})$, $\xi_m^i = \frac{\alpha^3}{3}$ by Lemma 2.2 where $i = 1$ or 2. By substituting $\alpha = a + b\sqrt[3]{m} + c\sqrt[3]{m^2}$ and considering the divisibility of y, z, y_1 and

z_1 by 3, one can explicitly see that 3 has to divide $(a^3 - mb^3)c$. If 3 divides c, then one easily shows that a and b are also divisible by 3. It leads to a contradiction that either x or x_1, and hence, either ξ_m or ξ_m^2 is divisible by 3.

The remaining possibility that 3 divides $a^3 - mb^3$ also leads to a contradiction when $m \equiv \pm 1 \bmod 3$, as consideration of $a^3 \equiv \pm b^3$ modulo 9 implies divisibility of x or x_1 by 3. Therefore, m is either a prime or a multiple of 3 when $3 \nmid h_m$. □

Corollary 2.4 *For any square-free composite number* $m \equiv 2, 4, 5$ *or* $7 (mod\ 9)$, *the class number of* $\mathbb{Q}(\sqrt[3]{m})$ *is divisible by* 3.

Now we prove Theorem 1.1 with the help of Lemma 2.3. Consider $K = \mathbb{Q}(\sqrt[3]{m})$ such that $3 \nmid h_m$. Suppose m is not a prime. By Lemma 2.3, we know that m must be divisible by 3. We want to show that $m = 3p$ for some prime p. Suppose m is divisible by two distinct primes p and q other than 3. Then, 3, p and q all ramify in K, and so does $3p$. By Lemma 2.2 there exist $\alpha,\ \beta \in K$ such that either $\frac{\alpha^3}{3} = \frac{\beta^3}{3p}$ or $(\frac{\alpha^3}{3})^2 = \frac{\beta^3}{3p}$ or $\frac{\alpha^3}{3} = (\frac{\beta^3}{3p})^2$. But these identities imply that a cube root of p or $9p$ belongs to $\mathbb{Q}(\sqrt[3]{m})$, which is not possible as $m \neq p, 9p$. Hence, the only possibility is $m = 3p$ where p is a prime. □

Now we can establish certain congruences for the fundamental unit of $K = \mathbb{Q}(\sqrt[3]{3p})$. By Lemma 2.2, one of ξ_m or ξ_m^2 can be expressed as $\frac{\alpha^3}{3}$ and as $\frac{\beta^3}{p}$ for some $\alpha,\ \beta \in K$. If $\xi_m = \frac{\beta^3}{p}$, then $y \equiv z \equiv 0 \pmod 3$ follows directly by expanding β^3. If $\xi_m^2 = \frac{\beta^3}{p}$, then $y_1 \equiv z_1 \equiv 0 \pmod 3$ follows similarly. In that case, one can easily deduce from (2.2) that $3 \mid xy$. If $3 \mid x$ then (2.2) leads to the contradiction that $3 \mid x_1$ as well. Therefore, $3 \mid y$ and by considering (2.2), $3 \mid z$.

Putting $\alpha = a + b\sqrt[3]{m} + c\sqrt[3]{m^2}$ in $\xi_m^j = \frac{\alpha^3}{3}$ and taking norm, we obtain

$$\begin{aligned} \mathrm{Norm}_{K/\mathbb{Q}}(\alpha)^3 &= 3^3 \mathrm{Norm}_{K/\mathbb{Q}}(\xi_m^j) \\ \Rightarrow \mathrm{Norm}_{K/\mathbb{Q}}(\alpha) &= a^3 + mb^3 + m^2c^3 - 3mabc = 3. \end{aligned} \tag{2.3}$$

Expansion of α^3 shows that either x or x_1 is $\frac{a^3+mb^3+m^2c^3+6mabc}{3}$ which equals $\frac{3+9mabc}{3}$ by (2.3). As m is divisible by 3, 3 must divide a. Hence either x or x_1 is $1 + 3mabc$. As $m = 3p$, we have either x or x_1 is congruent to 1 (mod $27p$). But $x_1 = x^2 + 2myz \equiv x^2 \pmod{27p}$ because $m = 3p$ and $y \equiv z \equiv 0 \pmod 3$. □

Examples: (i) The fundamental unit of $\mathbb{Q}(\sqrt[3]{3.2})$ is $109 + 60\sqrt{6} + 33\sqrt[3]{36}$ (see [5]), where $60 \equiv 0 \equiv 33 \bmod 3$ and $109^2 \equiv 1 \bmod 54$.
(ii) The fundamental unit of $\mathbb{Q}(\sqrt[3]{3.5})$ is $5401 + 2190\sqrt{6} + 888\sqrt[3]{36}$ (see [5]), where $2190 \equiv 0 \equiv 888 \bmod 3$ and $5401^2 \equiv 1 \bmod 135$.

Corollary 2.5 *Let* $\xi_m = x + y\sqrt[3]{3p} + z\sqrt[3]{9p^2}$ *be the fundamental unit of* $\mathbb{Q}(\sqrt[3]{3p})$. *If* $3 \nmid y$ *or* $3 \nmid z$ *or* $27p \nmid (x^2 - 1)$ *then the class number of* K *is divisible by* 3.

Examples: (i) The fundamental unit of $\mathbb{Q}(\sqrt[3]{3.7})$ is $1705 + 618\sqrt{21} + 224\sqrt[3]{441}$ (see [5]), where $1705^2 \not\equiv 1\ mod\ 27$. One can check that the class number is 3.

(ii) The fundamental unit of $\mathbb{Q}(\sqrt[3]{3.13})$ is $529 + 156\sqrt{6} + 46\sqrt[3]{36}$ (see [5]). Here $46 \not\equiv 0 \ mod\ 3$ and once can check that the class number is 6.

By similar arguments, we can obtain the following results concerning congruence relations satisfied by the fundamental unit of $K = \mathbb{Q}(\sqrt[3]{p})$ when $3 \nmid h_p$.

Proposition 2.6 *Let $\xi_m = x + y\sqrt[3]{p} + z\sqrt[3]{p^2}$ be the fundamental unit of $K = \mathbb{Q}(\sqrt[3]{p})$ where p is a prime $\equiv 4$ or $7 \ mod\ 9$. If $3 \nmid h_p$ then $x^2 \equiv 1 \ mod\ 3p$ and one of the following must hold:*
(i) $x \equiv 1 \ mod\ 3$ and $y + z \equiv 0 \ mod\ 3$.
(ii) $x \equiv 2 \equiv y \ mod\ 3$, $z \equiv 0 \ mod\ 3$.
(iii) $x \equiv 2 \equiv z \ mod\ 3$ and $y \equiv 0 \ mod\ 3$.

Proposition 2.7 *Let $\xi_m = x + y\sqrt[3]{p} + z\sqrt[3]{p^2}$ be the fundamental unit of $K = \mathbb{Q}(\sqrt[3]{p})$ where p is a prime $\equiv 2$ or $5 \ mod\ 9$. If $3 \nmid h_p$ then $x^2 \equiv 1 \ mod\ 3p$ and one of the following must hold:*
(i) $x \equiv 1 \ mod\ 3$ and $y - z \equiv 0 \ mod\ 3$.
(ii) $x \equiv 2 \equiv -y \ mod\ 3$, $z \equiv 0 \ mod\ 3$.
(iii) $x \equiv 2 \equiv z \ mod\ 3$ and $y \equiv 0 \ mod\ 3$.

The following theorem is now immediate from the two propositions above.

Theorem 2.8 *Let $\xi_m = x + y\sqrt[3]{p} + z\sqrt[3]{p^2}$ be the fundamental unit of $K = \mathbb{Q}(\sqrt[3]{p})$ where p is a prime $\not\equiv \pm 1 \ mod\ 9$. Suppose $3 \nmid h_p$. Then $x^2 \equiv 1 \ mod\ 3p$ and one of the following must hold.*
(i) $x \equiv 2 \ mod\ 3$, and 3 divides either y or z but not both.
(ii) $x \equiv 1 \ mod\ 3$ and $3 \mid (y + z)$ if $p \equiv 1 \ mod\ 3$; $3 \mid (y - z)$ if $p \equiv 2 \ mod\ 3$.

3 Real Quadratic Fields with Odd Class Number

In this section, we deduce certain congruence relations for the fundamental unit ξ_d of a real quadratic field $K = \mathbb{Q}(\sqrt{d})$ of odd class number. These congruence relations are stronger than similar ones proven in [6] by using 2-adic numbers, and in our earlier work [1] which used the same approach as here.

If $\mathbb{Q}(\sqrt{d})$ has odd class number then $d = p, 2p$ or pq where p and q denote distinct primes congruent to 3 modulo 4. When $d = pq \equiv 1$ mod 8 with $p \equiv 3 \equiv q$ mod 4, ξ_d still belongs to $\mathbb{Z}[\sqrt{d}]$ even though the ring of integers of $\mathbb{Q}(\sqrt{d})$ is larger than $\mathbb{Z}[\sqrt{d}]$. When $d = pq \equiv 5$ mod 8 with $p \equiv 3 \equiv q$ mod 4, ξ_d may not belong to $\mathbb{Z}[\sqrt{d}]$ but ξ_d^3 does.

First, we state an obvious analogue of Lemma 2.2 for the real quadratic case.

Lemma 3.1 (i) *If $d = p$ or $2p$ where p is a prime congruent to $3 \ mod\ 4$, then $2\xi_d = \alpha^2$ for some $\alpha \in \mathcal{O}_K$.*

(ii) If $d = pq$, where p and q are two distinct primes congruent to 3 mod 4, then $p\xi_d^j = \alpha^2$ for some $\alpha \in \mathcal{O}_K$, where $j = 1$ if $pq \equiv 1$ nod 8 and $j = 3$ if $pq \equiv 5$ mod 8.

The following result is classically known, but we can also deduce it from the second relation in Lemma 3.1.

Corollary 3.2 *If $K = \mathbb{Q}(\sqrt{d})$ is a real quadratic field with discriminant having at least three prime factors then the class number of K is even.*

Proof Suppose $K = \mathbb{Q}(\sqrt{d})$ is the real quadratic field under question. The given condition implies that at least three distinct rational primes p, q and r ramify in K. Hence, ideals generated by both p and pq ramify in K, but none of them is a square in K. If ξ_d denotes the fundamental unit of K and the class number of K is not divisible by 2, then we can write $\xi_d = \frac{\alpha^2}{p} = \frac{\beta^2}{pq}$ by Lemma 3.1. Then $\sqrt{q} \in K$, a contradiction. □

Theorem 3.3 *Let p and q be two distinct primes congruent to 3 modulo 4, and $x + y\sqrt{pq} \in \mathbb{Z}[\sqrt{pq}]$ be the fundamental unit ξ_{pq} of $K = \mathbb{Q}(\sqrt{pq})$ (or its cube when $pq \equiv 5$ mod 8). Without loss of generality, suppose p is a quadratic residue mod q. Then*
(i) $x \equiv 7$ mod 8, and $y \equiv 0$ mod 4.
(ii) $x - 1$ is divisible by $2q$ and $\frac{x-1}{2q}$ is the square of an odd integer.
(iii) $x + 1$ is divisible by $2p$ and $\frac{x+1}{2p}$ is the square of an even integer.

Suppose $d = pq$ where $p \equiv q \equiv 3$ mod 4. As the prime p ramifies in $\mathbb{Q}(\sqrt{pq})$, we have $x = \frac{a^2+pqb^2}{p}$ and $y = \frac{2ab}{p}$ by Lemma 3.1. We can obtain the following additional information about a and b.

Lemma 3.4 *(i) When $\left(\frac{p}{q}\right) = 1$, we have $a^2 - pqb^2 = p$.*

(ii) When $\left(\frac{p}{q}\right) = -1$, we have $a^2 - pqb^2 = -p$.

Proof By considering norm of the fundamental unit, we obtain $\left(\frac{a^2+pqb^2}{p}\right)^2 - pq\left(\frac{2ab}{p}\right)^2 = 1$, which implies $a^2 - pqb^2 = \pm p$. When $\left(\frac{p}{q}\right) = 1$, we have $\left(\frac{-p}{q}\right) = \left(\frac{-1}{q}\right) = -1$, hence $a^2 - db^2 \neq -p$. When $\left(\frac{p}{q}\right) = -1$, we have $\left(\frac{p}{q}\right) = -1$, hence $a^2 - db^2 \neq p$. □

Proof of Theorem 3.3 Consider the fundamental unit $\xi_{pq} = x + y\sqrt{pq}$ where p and q are primes congruent to 3 mod 4. By Lemmas 3.1 and 3.4, we have

$$\begin{aligned} &px = a^2 + pqb^2, \quad py = 2ab, \quad p = a^2 - pqb^2 \\ \Rightarrow \quad &p(x+1) = 2a^2, \quad x - 1 = 2qb^2. \end{aligned} \tag{3.1}$$

Considering $p = a^2 - pqb^2 \bmod 4$, we find that a is even and b is odd. In particular, $y \equiv 0 \bmod 4$. As $x - 1 = 2qb^2 \equiv 6 \bmod 8$, we have $x \equiv 7 \bmod 8$. Clearly, $x - 1$ is divisible by $2q$ and $\frac{x-1}{2q}$ is the square of an odd integer. Moreover, $x + 1$ is divisible by $2p$ and $\frac{x+1}{2p}$ is the square of an even integer. Theorem 3.3 follows. □

Examples: (i) The fundamental unit of $\mathbb{Q}(\sqrt{3.11})$ is $23 + 4\sqrt{33}$, where $23 \equiv 7 \bmod 8$, $4 \equiv 0 \bmod 4$, $\frac{23-1}{2.11} = 1^2$ and $\frac{23+1}{2.3} = 2^2$. Note that $\left(\frac{3}{11} = 1\right)$.
(ii) The fundamental unit of $\mathbb{Q}(\sqrt{11.19})$ is $46551 + 3220\sqrt{11.19}$, where $46551 \equiv 7 \bmod 8$, $3220 \equiv 0 \bmod 4$, $\frac{46551-1}{2.19} = 35^2$ and $\frac{46551+1}{2.11} = 46^2$.
Results similar to Theorem 3.3 can be obtained in the other two cases $d = p$ or $d = 2p$ when the quadratic field $\mathbb{Q}(\sqrt{d})$ has odd class number.

References

1. D. Chakraborty, A. Saikia, Congruence relations for the fundamental unit of a pure cubic field and its class number. J. Number Theory **166**, 76–84 (2016)
2. F. Gerth, Cubic fields whose class number are not divisible by 3. Illinois. J. Math. **20**(3), 486–493 (1976)
3. F. Lemmermeyer, Why the class number of $\mathbb{Q}(\sqrt[3]{11})$ is even? Math. Bohem. **138**, 149–163 (2013)
4. Stein. W. A, et al., *Sage Mathematics Software (Version 6.2 (2014-05-07))*, The Sage Development Team, http://www.sagemath.org
5. H. Wada, A table of fundamental units of purely cubic fields, in *Proceedings of the Japan Mathematical Society*, vol. 46 (1970), pp. 1135–1140
6. Z. Zhang, Q. Yue, Fundamental unit of real quadratic fields of odd class number. J. Number Theory **137**, 122–129 (2014)

The Charm of Units I, On the Kummer–Vandiver Conjecture. Extended Abstract

Preda Mihăilescu

Une paquerette, la midinette
Se balançait – dans un champ de blé ...
Une petite abeille, en plein soleil,
La féconda, trà la là la là.
To John Coates

1 Introduction

We introduce first the minimal notation that is needed for explaining the purpose and approach of this paper. Let p be an odd prime and $\tilde{\mathbb{K}} = \tilde{\mathbb{K}}_0 = \mathbb{Q}[\zeta]$ be the p-th cyclotomic field and $\tilde{\mathbb{K}}_n = \tilde{\mathbb{K}}[\zeta_{p^{n+1}}]$ be the intermediate fields of its cyclotomic $\mathbb{Z}_p$-extension. Let $G_n = \mathrm{Gal}(\tilde{\mathbb{K}}_n/\mathbb{Q})$ and $\Gamma_n = \mathrm{Gal}(\tilde{\mathbb{K}}_n/\tilde{\mathbb{K}}_0)$ be generated by the restriction of a topological generator $\tau \in \Gamma = \mathrm{Gal}(\tilde{\mathbb{K}}_\infty/\tilde{\mathbb{K}})$ of the galois group of the cyclotomic $\mathbb{Z}_p$-extension $\tilde{\mathbb{K}}_\infty$ of $\tilde{\mathbb{K}}$. Let $\nu_{n.a} : \zeta_{p^{n+1}} \mapsto \zeta_{p^{n+1}}^a$ be the automorphisms of the group G_n that send $\zeta_{p^{n+1}}$ to its a-th power, for $(a, p) = 1$. Thus $\sigma_c := \nu_{n.c^{p^n}}$ are lifts of G_0 to G_n. The maximal real subfield of $\tilde{\mathbb{K}}$ is $\mathbb{K} = \mathbb{Q}[\zeta + \overline{\zeta}]$—since most of our work takes place in the real subfields, we keep the notation $\mathbb{K}$ for this field and use the uncommon notation $\tilde{\mathbb{K}}$ for the full cyclotomic field, which has only a collateral significance in this paper.

If X is a finite abelian group, we denote by X_p its p—Sylow subgroup and for an arbitrary number field $\mathbf{K}$ we write $A(\mathbf{K}) = \mathcal{C}(\mathbf{K})_p$ for the p—part of the class group $\mathcal{C}(\mathbf{K})$. The units of this field are $E(\mathbf{K}) = \mathcal{O}^\times(\mathbf{K})$ and $E'(\mathbf{K})$ are the units of the ring of p-integers $\mathcal{O}(\mathbf{K})[1/p]$. The local units $U(\mathbf{K})$ are the invertible integers in the algebra $\mathbf{K}_p := \mathbf{K} \otimes_{\mathbb{Q}} \mathbb{Q}_p$.

P. Mihăilescu (✉)
Mathematisches Institut der Universität Göttingen, Göttingen, Germany
e-mail: preda@uni-math.gwdg.de

K. Chakraborty et al. (eds.), *Class Groups of Number Fields and Related Topics*,
https://doi.org/10.1007/978-981-15-1514-9_7

In a letter to Kronecker dating from 1849, Kummer explained his approach for proving Fermat's Last Theorem. The assumptions that

K1. The class group of $\mathbb{K}$ has trivial p-subgroup, or equivalently, $p \nmid |A(\mathbb{K})|$, and
K2. The p^2-primary units of $\mathbb{K}$ are global p-powers, or equivalently, $E(\mathbb{K}) \cap U^{p^2}(\mathbb{K}) \subset E^p(\mathbb{K})$,

play a central role in this approach. We know that Kummer succeeded later to prove the second case of Fermat's Last Theorem (FLT) on base of these very assumptions. Four years later, writing to the same Kronecker, Kummer still mentioned K1 as *ein noch zu beweisender Satz*,[1] which raises some problems. Kummer never succeeded to prove the assumption, and in the early decades of the twentieth century, Vandiver, who pursued the work of Kummer on Fermat, used this assumption in several of his results. The assumption K1 is currently referred to as Conjecture of Kummer–Vandiver. It can be found together with some of its more important consequences in the textbooks on cyclotomy of Lang [4] and Washington [7]. Vandiver indicated also an approach to the study of the First Case of FLT, which is based on the Kummer–Vandiver Conjecture. The approach was made rigorous by Sitaraman in 1993 [5], and it states that the First Case holds, if along with K1, the condition K3. holds:

K3. The p-part $A(\tilde{\mathbb{K}})$ has exponent p.

The conjecture of Kummer–Vandiver for the p-th cyclotomic field is related to the K-theory of $\mathbb{Q}$. In particular, the $(p-1-2n)$-th component of $A(\mathbb{K})$ vanishes if $K^{4n}(\mathbb{Q}) \otimes \mathbb{Z}_p = 0$. Grzegorz Banaszak and Vojtek Gajda proved this result in 1990 [1] and deduced from the fact that $K^{4n}(\mathbb{Q})$ is a torsion group, that for any $n > 0$, the $(p-1-2n)$-th component of $A(\mathbb{K})$ vanishes for all sufficiently large p. Using the proof of $K^4(\mathbb{Q}) = 0$ given by Vojevodski, Kurihara deduced in 1993 [3] that $A^{(p-3)}(\mathbb{K}) = 0$ for all p. This result was improved by Greenberg and then Soulé in [6].

In this paper we shall show.

Theorem 1 *Suppose that the groups A_n^+ are uniformly bounded. Then they are trivial.*

In other words, we show that Greenberg's λ-conjecture, stating the finiteness of the projective limits of p-parts of class groups in the cyclotomic $\mathbb{Z}_p$-extension of a totally real field, implies Kummer–Vandiver. The former conjecture is proved in the general case, in a separate paper.

1.1 Plan of the Proof

Greenberg's λ-conjecture states that for arbitrary totally real fields $\mathbf{K}$, the groups $A_\infty(\mathbf{K})$ are finite. In particular, when $\mathbf{K} = \mathbb{K}$, this implies that $|A(\mathbb{K}_n)|$ are uniformly

[1] A yet to prove theorem.

bounded when $n \to \infty$. We shall assume the truth of the Greenberg Conjecture, which shall be proved for arbitrary CM fields, in a separate paper. We thus assume that there is some $M > 0$ such that $|A_n| < M$ for all $n \in \mathbb{N}$, but $A_n \neq \{1\}$. We denote by $\mathbb{H}_n$ the maximal p-abelian unramified extension of $\mathbb{K}_n$, which is herewith a finite extension of bounded size, for $n \to \infty$.

We fix some real extension $\mathbb{L}/\mathbb{K}$ which is unramified, cyclic of degree p and Galois over $\mathbb{Q}$, and let $\mathbb{L}_n = \mathbb{K}_n \cdot \mathbb{L} \subset \mathbb{H}_n$; we fix a generator $\nu \in \text{Gal}(\mathbb{L}/\mathbb{K})$—since $\mathbb{L}/\mathbb{K}$ is real and unramified, its existence is equivalent to the failure of the Kummer–Vandiver conjecture. The structure of the local and global units of $\mathbb{L}_n$ is intimately connected to the simpler structure of the local units of $\mathbb{K}_n$, in which there is only one prime above p. Moreover, capitulation happens both in $\mathbb{L}_n/\mathbb{K}_n$ for all n, and also, for sufficiently large n, in the vertical extensions $\mathbb{K}_{n+1}/\mathbb{K}_n$. By a detailed study of global and local units in $\mathbb{L}_n$, we find incompatibility between these two types of capitulation. In this investigation, such classical results as Hilbert's Theorems 90–94, together with some application of local class field theory and Kummer theory join together in order to reveal a contradiction.

More precisely, we show that for each n there is some *singular* unit $\delta_n \in E(\mathbb{L}_n)$ which has norm one and does not lay in the augmentation ideal of the units:

$$\mathbf{N}_{\mathbb{L}/\mathbb{K}}(\delta_n) = 1, \qquad \delta_n \notin E(\mathbb{L}_n)^{\nu-1}.$$

There also is a local *singular $\mathbb{Z}_p$-module S_n*—defined in Definition 4 below—, contained in the local units of $\mathbb{L}_n$; for sufficiently large n, the projections of δ_n to S_n define a submodule of finite index

$$p^{\ell_n} = [S_n : \delta_n^{\mathbb{Z}_p[s]}]$$

in the singular module. It turns out that $\ell_n > 0$ but the index decreases by one in each extension $\mathbb{K}_{n+1}/\mathbb{K}_n$, for sufficiently large n. This is the contradiction which proves our Theorem. The plan is only descriptive at this point, but the notions introduced will become clear and shall be formally defined along the proof below.

1.2 Notations and Auxiliary Results

For any CM field $\mathbf{K}$ we write $\jmath$ for the image of complex conjugation in $\text{Aut}(\mathbf{K}/\mathbb{Q})$. The $\mathbb{Z}_p$-extension of $\mathbb{Q}$ is $\mathbb{Q}_\infty$ and $\mathbb{Q}_n \subset \mathbb{Q}_\infty$ are the subfields of degree p^n over $\mathbb{Q}$. Accordingly, if $\mathbf{K}$ is an arbitrary number field, its cyclotomic $\mathbb{Z}_p$-extension is $\mathbf{K} \cdot \mathbb{Q}_\infty$ and $\mathbf{K}_n = \mathbf{K} \cdot \mathbb{Q}_n$. We fix a p-th root of unity ζ and write $\tilde{\mathbb{K}}_n = \mathbb{Q}_n[\zeta] = \mathbb{Q}[\zeta_{p^{n+1}}]$. Then $\tilde{\mathbb{K}} = \tilde{\mathbb{K}}_0$ and the galois group of $\tilde{\mathbb{K}}_0$ is

$$\begin{aligned} G = G_0 = {} & \text{Gal}(\tilde{\mathbb{K}}_0/\mathbb{Q}) \\ = {} & \{\sigma_a \ : \ a = 1, 2, \ldots p-1, \quad \zeta \mapsto \zeta^a\} \cong (\mathbb{Z}/p \cdot \mathbb{Z})^*. \end{aligned}$$

The group G lifts to $G_n = \mathrm{Gal}(\tilde{\mathbb{K}}_n/\mathbb{Q}) \cong \Gamma_n \times G$, with $\Gamma_n = \mathrm{Gal}(\mathbb{K}_n/\mathbb{K}) \cong \mathrm{Gal}(\mathbb{Q}_{n-1}/\mathbb{Q})$. The groups G_n are cyclic and we may denote by $\tilde{\nu}_a \in G_n$ the automorphisms given by $\tilde{\nu}_a : \zeta_{p^n} \rightarrow \zeta_{p^n}^a$; in this notation, we identify σ_a with its lift $\tilde{\sigma}_a = \tilde{\nu}_{a^{p^{n-1}}} \in G_n$. The maximal real subfields of $\tilde{\mathbb{K}}_n$ are $\mathbb{K}_n = \mathbb{Q}_n[\zeta + \overline{\zeta}]$. The ramified primes above p are

$$\tilde{\wp}_n = (\tilde{\lambda}_n), \quad \wp_n = (\lambda_n), \qquad \text{with}$$
$$\tilde{\lambda}_n = \zeta_{p^{n+1}} - \overline{\zeta}_{p^{n+1}}, \quad \lambda_n = \tilde{\lambda}_n^2.$$

The cyclotomic units are

$$C(\mathbb{K}_n) = \tilde{\lambda}_n^{\mathbb{Z}[G_n]} \bigcap E(\mathbb{K}_n) \subset E(\mathbb{K}_n);$$

these modules are generated by $\eta_n = \tilde{\lambda}_n^{\sigma-1}$, as a $\mathbb{Z}[G_n]$-module, up to possible index 2. For $n = 0$, we may also write $\eta = \eta_0$.

For $\mathbf{R} \in \{\mathbb{F}_p, \mathbb{Z}_p, \mathbb{Z}/(p^m \cdot \mathbb{Z})\}$, we describe the group rings $\mathbf{R}[G]$ by means of the Teichmüller character, induced by unique $p-1$-th roots of unity, verifying:

$$\varpi(\sigma_a) \in \mathbf{R}, \quad \varpi(\sigma_a) \cong a \bmod p\mathbf{R}.$$

Herewith, the *orthogonal idempotents* $e_k \in \mathbf{R}[G]$ are

$$e_k = \frac{1}{p-1} \sum_{a=1}^{p-1} \varpi^k(\sigma_a) \cdot \sigma_a^{-1}, \tag{1}$$

and some of the most usual properties (see also [7], p. 100) are

$$\sum_{k=0}^{p-2} e_k = 1, \quad \text{and} \quad (\sigma_c - \varpi^k(\sigma_c))e_k = 0. \tag{2}$$

We shall use the notation e_k quite freely—at times it denotes the p-adic idempotents for $\mathbf{R} = \mathbb{Z}_p$, but we may also choose some $t_k \in \mathbb{Z}[G]$ which approximates e_k to some power p^M, and denote these also by e_k, when there is no place for confusion; in all cases, the orthogonality relations hold modulo p^M, and the precise signification of e_k will be made clear in the context.

The local units $U(\mathbf{K})$ are defined as follows. Let $P = \{\mathfrak{P}_j : j = 1, 2, \ldots, s\}$ be the primes of $\mathbf{K}$ above p. Then the p-idèles of $\mathbb{K}$ are

$$\mathbf{K}_p := \mathbf{K} \otimes_{\mathbb{Q}} \mathbb{Q}_p \cong \prod_{j=1}^{s} \mathbf{K}_{\mathfrak{P}_j},$$

where $\mathbf{K}_{\mathfrak{P}}$ is the completion of $\mathbf{K}$ at the prime $\mathfrak{P}$ and $\mathbb{Q}_p$ embeds diagonally in $\mathbf{K}_p$. The local units are then defined as the products of the units in the individual completions, and their subgroup U' is the kernel of the global norm $N_p : \mathbf{K}_p \to \mathbb{Q}_p$:

$$U(\mathbf{K}) := \prod_{\mathfrak{P} \in P} \mathcal{O}\left(\mathbf{K}_{\mathfrak{P}}\right), \quad U'(\mathbf{K}) = \operatorname{Ker}(N_p : U(\mathbf{K}) \to \mathbb{Z}_p). \tag{3}$$

By definition, $\mathbf{K}$ is dense in $\mathbf{K}_p$ in the p-adic product topology, so any $x \in \mathbf{K}_p$ is approximated by sequences $(x_n)_{n \in \mathbb{N}} \subset \mathbf{K}$ with $x - x_n \in U(\mathbf{K})^{p^n}$. The p-adic closure of $E(\mathbf{K})$ is

$$\overline{E(\mathbf{K})} = \bigcap_{N=1}^{\infty} \left(E(\mathbf{K}) \cdot U(\mathbf{K})^{p^N}\right) \subset U'(\mathbf{K}),$$

and $E(\mathbf{K})$ embeds diagonally in the completion.

Definition 1 If $\mathbf{L}/\mathbf{K}$ is a cyclic extension of degree p of the field $\mathbf{K}$, with group $X = \langle \nu \rangle$, and $s = \nu - 1$ we let

$$S = \{f \in \mathbb{Z}[s] \ : \ \deg(f) < p - 1\}$$

and denote a set of units $\mathcal{H} = \{H_0; H_1, H_2, \ldots, H_k\} \subset E(\mathbf{L})$ as *fundamental set of Hilbert relative units*, and $\underline{\mathcal{H}} = \mathcal{H} \setminus \{H_0\}$ a pre-fundamental set of Hilbert relative units, if the following conditions are fulfilled:

1. The unit H_0 verifies $\mathcal{N}(H_0) = 1$ but $H_0 \notin E(\mathbf{L})^{(s,p)}$,
2. The set
$$\mathbf{H}' = \mathbf{H}(\underline{\mathcal{H}}) = \left\{\prod_{i=1}^{k} H_i^{f_i} \ : \ f_i \in S\right\} \subset E(\mathbf{L}),$$
the *restricted* $\mathbb{Z}[s]$*-hull of* $\mathcal{H}$, verifies $\mathcal{N}(\mathbf{H}') = \mathcal{N}(E(\mathbf{L}))$.
3. The units in $\mathcal{H}^S := \{\nu^j(H_i) \ : \ i = 0, 1, \ldots, k; \ j = 0, 1, \ldots, p-2\}$ are $\mathbb{Z}$-independent and together with a fundamental set of units for $E(\mathbb{K})$, they build a fundamental set of units for $E(\mathbb{L})$.

By a slight abuse of language, we shall also say $\mathcal{H}$ are a fundamental set of relative units, if the index $[E(\mathbf{L}) : E(\mathbf{K}) \cdot \mathbf{H}]$ is coprime to p.

We introduce some additional notation connected to the group ring $\mathbb{Z}[X] = \mathbb{Z}[s]$; here, $\mathcal{N} = 1 + \nu + \cdots + \nu^{p-1}$ is the associated *norm*, and its *derivative* is defined by

$$\mathcal{N} = \frac{(s+1)^p - 1}{s} = p + s\mathcal{N}', \tag{4}$$

$$\mathcal{N}' := s^{p-2} + p \cdot \left(\sum_{i=2}^{p-1} \frac{\binom{p}{i}}{p} \cdot s^{i-2}\right).$$

We also note the following relations which will be useful in the sequel. The proof is a simple application of the developments in (4) and is left to the reader.

$$s \cdot \mathcal{N} = 0 = s^p + psv^{-1}(s),$$
$$v(s) = 1 + O(s) \Rightarrow ps = -v(s)s^p, \tag{5}$$
$$\mathcal{N} = s^{p-2} + pu_1(s), \quad u_1(s) = \frac{p-1}{2} + O(s) \in (\mathbb{Z}_p[s])^\times. \tag{6}$$

We assume in this paper that the Kummer–Vandiver conjecture is false and, concretely, there is an integer $2m \in \{2, 4, \ldots, p-3\}$, such that $e_{2m}A \neq \{1\}$. We may accordingly choose the field $\mathbb{L}$ mentioned above, such that $\mathrm{Gal}(\mathbb{L}/\mathbb{K})$ is a quotient of $\mathrm{Gal}(\mathbb{H}_0/\mathbb{K})^{e_{2m}}$. Then $\mathbb{L}$ is galois over $\mathbb{Q}$, as a consequence of the fact that $e_{2m}\mathbb{Z}_p[G_0]$ has rank 1.

We adopt the notation to the case of $X = \mathrm{Gal}(\mathbb{L}/\mathbb{K})$, a group generated by ν; the augmentation is generated by $s = \nu - 1$ and the norm is $\mathcal{N} = \mathbf{N}_{\mathbb{L}/\mathbb{K}}$. Thus $X \cong \mathrm{Gal}(\mathbb{L}_n/\mathbb{K}_n) \cong \mathrm{Gal}(\mathbb{L}/\mathbb{K})$ does not depend on n. We denote by Y_n the groups

$$Y_n = \mathrm{Gal}(\mathbb{L}_n/\mathbb{Q}) = G_n \ltimes X.$$

The maximal p-abelian, p-ramified extension of $\tilde{\mathbb{K}}_n$ is $\mathbb{H}_n$ and $\Omega_{E.n} = \tilde{\mathbb{K}}_n\left[E(\tilde{\mathbb{K}}_n)^{1/p^n}\right]$. In the injective limit we write

$$\mathbb{H}_\infty = \bigcup_n \mathbb{H}_n; \quad \Omega_E = \bigcup_n \Omega_{E.n}.$$

Complex conjugation $\jmath$ acts by conjugation on the galois groups of these extensions inducing natural decomposition in plus and minus parts: writing $Z_n = \mathrm{Gal}(\mathbb{H}_n/\tilde{\mathbb{K}}_n)$, $Z = \varprojlim_n Z_n$, we have

$$Z^- = (1-\jmath)Z, \quad Z^+ = (1+\jmath)Z; \qquad \mathbb{H}_\infty^- = \mathbb{H}_\infty^{Z^+}, \quad \mathbb{H}_\infty^+ = \mathbb{H}_\infty^{Z^-},$$

and similar for the various other fields introduced above.

1.3 Hilbert's Theorems on Class Fields

In this section, we recall the most important of the Theorems 90–94 on class fields in Hilbert's Zahlbericht [2]. Hilbert considers an arbitrary cyclic galois extension $\mathbf{L}/\mathbf{K}$ of degree p, with galois group $X = \langle \nu \rangle$ and Dirichlet number $r = r(\mathbf{K}) = r_1 + r_2 - 1$. Then the Dirichlet number of $\mathbf{L}$ is $R = r(\mathbf{L}) = p(r+1) - 1$. The Theorem 90 implies the vanishing of $\widehat{H}^0(X, \mathbf{L}^\times)$: it states that for $x \in \mathbf{L}^\times$ with $\mathcal{N}(x) = 1$, there is a $y \in \mathbf{L}^\times$ such that $x = y^{\nu-1}$. The Theorem 91 is less often quoted or used, but it plays an important role in our approach. Here is its slightly modified statement that we shall use:

Fact 1 (Hilbert's Theorem 91) *Let* $\mathbf{L}/\mathbf{K}$ *be an extension like in definition (1) and let* $r = r_1 + r_2 - 1$ *be the Dirichlet rank of* $E(\mathbf{K})$. *Then there is a fundamental set of Hilbert relative units* $\mathcal{H} = \{H_i \ : \ i = 0, 1, \ldots r\} \subset E(\mathbf{L})$, *such that* $H_0 \notin E(\mathbf{L})^{(p,s)}$ *and* $\mathcal{N}(H_0) = 1$. *Moreover, if* $\mathbf{H}'$ *is the restricted* $\mathbb{Z}[s]$*-hull of* $\mathcal{H}' = \mathcal{H} \setminus \{H_0\}$, *then the index* $[\mathcal{N}(E(\mathbf{L})) : \mathcal{N}(\mathbf{H}')]$ *is coprime to* p.

In the case that $\mathbf{L}/\mathbf{K}$ is unramified, the unit H_0 gives raise to capitulation, according to the celebrated Theorem 94

Fact 2 (Hilbert's Theorem 94) *If* $\mathbf{L}/\mathbf{K}$ *is unramified and* $\delta \in E(\mathbf{L}) \setminus E(\mathbf{L})^s$ *has norm* $\mathcal{N}(\delta) = 1$, *then there is an ideal* $\mathfrak{A} \subset \mathbf{K}$, *which is not principal, but lifts to a principal ideal* $(\gamma) = \mathfrak{A} \cdot \mathcal{O}(\mathbf{L})$, *such that* $\gamma^s = \delta$.

Proof By Theorem 90, there is some $\gamma \in \mathbf{L}$ such that $\delta = \gamma^s$, so the ideal (γ) is fixed by $\mathrm{Gal}(\mathbf{L}/\mathbf{K})$. Since the extension is unramified, it must be the lift of an ideal from $\mathfrak{A} \subset \mathcal{O}(\mathbf{K})$. If this ideal is principal, say $\mathfrak{A} = (\alpha)$, then there is some unit $d \in E(\mathbf{L})$ such that $\gamma = d\alpha$ and thus $\delta = (d\alpha)^s = d^s \in E(\mathbf{L})^s$, in contradiction with the choice of δ. We thus obtain a map $\widehat{H}^1(\mathrm{Gal}(\mathbf{L}/\mathbf{K}), E(\mathbf{L})) \to A(\mathbb{K})[p]$, by which $\delta E(\mathbf{L})^s$ has a nontrivial image. □

2 Primes and Local Units

We start with a first glance at the primes above p in our fields of interest.

2.1 Primes Above p

Let $n \geq 0$ and $\wp_n = (\lambda_n) = (\xi - \overline{\xi})^{1+\jmath}$ be the unique totally ramified prime above p in $\mathbb{K}_n$—with $\xi = \zeta_{p^n}$. Since $\mathbb{L}_n$ is unramified, the principal ideal theorem of Hilbert class fields implies that $\wp_n$ is split in $\mathbb{L}_n/\mathbb{K}_n$. *We assume in this paper, for simplicity, that the primes above* $\wp_n$ *are principal ideals. The general case is marginally more complex and will be treated in the final version of the paper.*

Let $X = \mathrm{Gal}(\mathbb{L}/\mathbb{K})$ be generated by ν and $s = \nu - 1, \mathcal{N} = \mathbf{N}_{\mathbb{L}/\mathbb{K}}$; since $Y :=$ $\mathrm{Gal}(\mathbb{L}/\mathbb{Q})$ is galois, it follows that the canonical lift of σ to $\mathrm{Gal}(\mathbb{K}/\mathbb{Q})$, acting by conjugation on X, fixes this group. Thus $e_{2m}X = X$ for some $m \in \{0, 1, \ldots, d\}$. The definition of m implies that $\nu^{\sigma_c} = \nu^{\varpi^{2m}(\sigma_c)}$, where conjugation acts via $\nu^{\sigma_c} = \sigma_c \nu \sigma_c^{-1}$ ([7], Sect. 10.2). We obtain

$$\sigma_c \circ \nu = \nu^{\varpi^{2m}(\sigma_c)} \circ \sigma_c, \quad \nu \circ \sigma_c = \sigma_c \circ \nu^{\varpi^{-2m}(\sigma_c)}. \tag{7}$$

We fix now $\mathfrak{p}_n \subset \mathbb{L}_n$, one of the primes above p, and let $\pi_n = \pi(\mathbb{L}_n) \in \mathbb{L}_n$ generate this principal ideal. We claim that π_n can be chosen such that $\mathcal{N}(\pi_n) = \lambda_n$; as ideals, we have $\mathcal{N}(\mathfrak{p}_n) = \wp_n$, so there is a unit $\varepsilon \in E(\mathbb{K}_n)$ such that $\mathcal{N}(\pi_n) = \varepsilon \cdot \lambda_n$. We may assume that $\varepsilon \in C_n$, the cyclotomic units of $\mathbb{K}_n$: if not, there is a minimal $j > 0$ such that $\varepsilon^{p^j} \in C_n$, and then $\mathcal{N}(\pi_n/\varepsilon^{p^{j-1}})/\lambda_n \in C_n$, as claimed. Since $C_n = \lambda_n^{\mathbb{Z}[\sigma,\tau]} \cap E(\mathbb{K}_n)$ (see, e.g., [7], Proposition 8.1), there is a group ring element

$$\theta = \sum_{i=0}^{N} c_i(\sigma - 1)T^i \in \Lambda; c_i \in \mathbb{Z}[X], \quad c_0(0) = 0,$$

such that $\varepsilon = \lambda_n^{\theta}$ and thus $\mathcal{N}(\pi_n) = \delta^{1+\theta}$. Note that for $A \in \mathbb{Z}[X]$,

$$(1 + (\sigma - 1)A(\sigma - 1))^p \equiv 1 + (\sigma - 1)A(\sigma - 1) \bmod p\mathbb{Z}[\sigma], \quad \text{so}$$
$$(1 + \theta) \cdot (1 + c_0(\sigma - 1)^{p-2}) \equiv 1 + O(T) \bmod p(\sigma - 1, T).$$

Moreover, $T^{p^n} \equiv 0 \bmod p((\sigma - 1), T)$, so combining with the previous identity, we see there is some $\theta_1 \in \mathbb{Z}[\sigma, \tau]$, such that $(1 + \theta)(1 + \theta_1) \equiv 1 \bmod p(\sigma - 1, T)$, while $\delta_1 := \pi_n^{\theta_1} \in E(\mathbb{L}_n)$. Consequently $\mathcal{N}(\delta_1\pi_n) = \lambda^{1+p\cdot\theta_2}$, for some $\theta_2 \in (\sigma - 1, T)$. But then $\lambda_n^{p\theta_2} = \varepsilon_1^p \in C_n^p$, and finally $\mathcal{N}(\pi_n\delta_1/\varepsilon_1) = \lambda_n$, which confirms the claim.

Since p is totally ramified in $\mathbb{K}_n/\mathbb{Q}$, it follows that the decomposition group $D(\mathfrak{p}_n) \subset \mathrm{Gal}(\mathbb{L}_n/\mathbb{Q})$ of $\mathfrak{p}_n$ is equal to its inertia group and $D(\mathfrak{p}_n) \cong \mathrm{Gal}(\mathbb{K}_n/\mathbb{Q})$. We can thus choose lifts $\tilde{\sigma}, \tilde{\tau} \in D(\mathfrak{p}_n)$ of the generators $\sigma, \tau \in \mathrm{Gal}(\mathbb{K}_n/\mathbb{Q})$ of the cyclic subgroups of order $\frac{p-1}{2}$ and p^n, respectively. In fact, τ commutes with X, so the lift $\tilde{\tau}$ is canonic and shall be identified to τ in the sequel. We also can fix $\nu \in X$ a generator verifying $\nu\omega = \zeta_p\omega$, for $\mathbb{L} = \mathbb{K}[\omega] = \mathbb{K}[\alpha^{1/p}]$, as a Kummer extension. We shall prove in Lemma 4 in Sect. 3, that the primes $\mathfrak{p}_n$ are principal, say $\mathfrak{p}_n = (\pi(\mathbb{L}_n))$ and one can choose π_n such that $\mathcal{N}(\pi_n) = \lambda_n$. Using the lifts $\tilde{\sigma}$, we note that

$$\mu_0(\mathbb{L}_n) := \pi(\mathbb{L}_n)^{\tau-1}, \quad \mu_1(\mathbb{L}_n) := \pi(\mathbb{L}_n)^{\tilde{\sigma}-1} \tag{8}$$

are units: this follows from the choice of the lifts, in the inertia group of π_n.

Definition 2 The units $\mu_0(\mathbb{L}_n) = \pi_n^{\tilde{\tau}-1}$, $\mu_1(\mathbb{L}_n) = \pi_n^{\tilde{\sigma}-1}$ are *metacyclotomic* units of $\mathbb{L}_n$ and they have the property that $\mathcal{N}(\mu_0), \mathcal{N}(\mu_1)$ generate C_n as a $\mathbb{Z}[G_n]$-module.

2.2 *Local Units in Fields and p-idèles*

We keep the notation introduced above, in particular $X = \langle \nu \rangle = \mathrm{Gal}(\mathbb{L}/\mathbb{K}) = \mathrm{Gal}(\mathbb{L}_n/\mathbb{K}_n)$ and both $\mathbb{L}_n$ and $\mathbb{K}_n$ are totally real, while $\tilde{\mathbb{K}}_n = \mathbb{K}_n[\zeta]$. We have fixed a principal prime $\mathfrak{p}_n = (\pi_n) \subset \mathbb{L}_n$ with $\mathcal{N}(\pi_n) = \lambda_n$ and defined lifts of σ, τ in its decomposition group, introducing by their means the metacyclotomic units in Definition 2.

We review the structure of local units in cyclotomic fields first, and then derive the structure of $U(\mathbb{L}_n)$. Subsequently, we focus our attention on the first level $\mathbb{K} = \mathbb{K}_0$ and show that the metacyclotomic units generate, together with an additional, *singular* unit δ_0, a system of Hilbert relative units, in the sense of Definition 1. The same system is also independent in the completion $\overline{E}(\mathbb{L})$, which proves the—ephemeral—result that the Leopoldt Conjecture holds in $\mathbb{L}$. This results deduced in detail at the level $\mathbb{K}_0$ generalizes quite easily for the units of $\mathbb{L}_n$; thus we define in particular a series of *singular* units $\delta_n \in E(\mathbb{L}_n)$. These are connected to capitulation and must change with n. Their projections to an invariant *singular* submodule $S_0 \subset U'(\mathbb{L}_n)$ generate under the action of $\mathbb{Z}_p[s]$ the projection of $E(\mathbb{L}_n)$ to the singular module. This fact will be led in the last chapter to a contradiction. In the present chapter, we prepare the necessary facts on units for the final proof.

2.2.1 Local Units in the Cyclotomic Extension

We recall first the structure of $U(\tilde{\mathbb{K}}_n)$; the idempotents e_k act naturally on $U(\tilde{\mathbb{K}}_n)$ inducing a direct product

$$U(\tilde{\mathbb{K}}_n) = \prod_{i=0}^{p-2} U^{(k)}(\tilde{\mathbb{K}}_n), \quad U^{(k)}(\tilde{\mathbb{K}}_n) = e_k U(\tilde{\mathbb{K}}_n).$$

Lang proves in [4], Sect. 6.2, that $U^{(k)}(\tilde{\mathbb{K}}_n)$ are cyclic Λ-modules for $k \notin \{0, 1\}$. One verifies from the definitions, that we also have $U^{(k)}(\tilde{\mathbb{K}}_n) \in U'(\tilde{\mathbb{K}}_n)$ for $k > 0$. In the case of $k = 1$, there exists a torsion submodule which obstructs cyclicity; this does not impact upon the cyclicity of $U(\mathbb{K}_n)$, which is the union of the *even* components of $U(\tilde{\mathbb{K}}_n)$. For $k = 0$, we have

Lemma 1 *The group $U'^{(0)}(\mathbb{K}_n)$ is Λ-cyclic, generated by $\Delta_n = \rho_n^T$, with $\rho_n = \mathbf{N}_{\mathbb{K}_n/\mathbb{Q}_n}(1 - \zeta_n)$ with ζ_n a primitive p^{n+1}-th root of unity and $\mathbb{Q}_n \subset \mathbb{K}_n$ the subfield of the cyclotomic $\mathbb{Z}_p$-extension $\mathbb{Q}_\infty/\mathbb{Q}$ that is contained in $\mathbb{K}_n$. Moreover,*

$$U^{(0)}(\mathbb{K}_n) = \mathbb{Z}_p^\times \times \Delta_n^\Lambda.$$

In particular, for $n = 0$, we have $U'(\mathbb{Q}_n) = \{1\}$ and $U(\mathbb{Q}_n) = \mathbb{Z}_p^\times$.

Proof It is known ([7], Proposition 8.1), that Δ_n generates the cyclotomic units of $\mathbb{Q}_n$, and since the class group of this field has vanishing p-part, the analytic class number formula applied to $\mathbb{Q}_n$ implies that $p \nmid [E(\mathbb{Q}_n) : C(\mathbb{Q}_n)]$, so Δ_n generates $E(\mathbb{Q}_n)$ as a $\mathbb{Z}[\tau]$-module, up to finite index, coprime to p. Consequently, Δ_n generates the cyclic submodule $\overline{E(\mathbb{Q}_n)} \subset U'(\mathbb{Q}_n)$. Moreover, $U'(\mathbb{Q}_n)/\overline{E(\mathbb{Q}_n)} \cong \mathrm{Gal}(R_n/\mathbb{Q}_n)$, where R_n is the maximal p-ramified p-abelian extension of $\mathbb{Q}_n$ which is fixed by (a lift of) Γ. By (Leopoldt) reflection, the radical in $\tilde{\mathbb{K}}_n$ of an extension of degree p, which is

Kummer and p-ramified over $\mathbb{Q}_n$ lays in $(\mathbb{K}_n^\times)^{e_1}$ and it is either a unit or generates a principal ideal which is the p-th power of a class in $e_1 A(\mathbb{K}_n)$. This component is trivial by Herbrand's Theorem [7], p. 101, and the only units in the first component are the roots of unity. Consequently, the only possible unramified abelian p-extension is $\mathbb{Q}_{n+1}$ itself. It follows that $U'(\mathbb{Q}_n)/\overline{E(\mathbb{Q}_n)} = 1$ and it is Λ- cyclic by the above proof. For $n = 0$, we have $\rho_0 = p$ and thus $\Delta_0 = 1$. Since $U(\mathbb{Q}_n)/U'(\mathbb{Q}_n) \cong \mathbb{Z}_p \subset U(\mathbb{Q}_n)$, the proof is complete. □

The following definition introduces generators of the Λ-cyclic components of $U'(\mathbb{K}_n)$:

Definition 3 For fixed n, we let $\xi_{n,k} \in U(\mathbb{K}_n)$ be generators of $U^{(k)}(\tilde{\mathbb{K}}_n)$ for $k \neq 0, 1$ and $\xi_{n,0} = \Delta_n$. Let $\xi_n = \prod_{k=0}^{d} \xi_{n,2k}$ and $\upsilon = 1 + p \in \mathbb{Z}_p$. If the value of n is fixed in the context, then we simply write $\xi, \xi^{(k)}$, etc. By definition,

$$U'(\mathbb{K}_n) = \xi_n^{\Lambda[G_0]}, \quad U'^{(2k)}(\mathbb{K}_n) = (\xi_n^{(2k)})^{\Lambda}, \quad \mathbb{Z}_p^\times = \upsilon^{\mathbb{Z}_p}, \tag{9}$$

2.2.2 Local Units in $\mathbb{L}_n$

The groups $Y_n = \text{Gal}(\mathbb{L}_n/\mathbb{Q})$ are semi-direct products $Y_n = G_n \ltimes X_n$, and we define the group rings $R_n := \mathbb{Z}_p[Y_n]$ or $R'_n = R_n/(\mathcal{N})$. Note that $sR_n = R_n s$, as follows from (7). Since Y_n is non commutative, caution is needed when using exponential notation, since when transcribing to functorial notation, the operations are reversed: $\alpha(\beta x) = x^{\beta\alpha}$, for $\alpha, \beta \in R_n$. We thus only use functional notation for the group ring operations, with the possible exception of the action of elements like s. An important consequence of the fact that sR_n is a bimodule, is the fact that there is a bijection $R_n/(sR_n) \to (sR_n)\backslash R_n$, so both factors are simultaneously finite or infinite.

Since $sR_n = R_n s$, and in view of (7), we deduce that every $\theta \in R_n$ can be represented by some polynomials $f_k, g_k \in \mathbb{Z}_p[s, T]$ as follows:

$$\theta = \sum_{g \in G_n} t_g g = \sum_{k=0}^{d} f_k(s, T) e_{2k} = \sum_{k=0}^{d} e_{2k} g_k(s, T), \tag{10}$$

where the idempotents e_{2k} are taken with respect to the fixed lift $\tilde{\sigma} \in X_n$ of $\sigma \in G_0$. Note also, that by choice of $\mathbb{L}_n$, the group $\text{Gal}(\mathbb{L}_n/\mathbb{L}) \cong \text{Gal}(\mathbb{K}_n/\mathbb{K}) \cong \Gamma/\Gamma^{p^n}$ commutes with ν.

With respect to the primes $\mathfrak{p}_n = (\pi_n)$, we have the p-idèles $(\mathbb{L}_n)_p = \prod_{i=0}^{p-1} (\mathbb{L}_n)_{\nu^i(\mathfrak{p})}$, and there are projections $\iota_i : (\mathbb{L}_n)_p \to (\mathbb{L}_n)_{\nu^i(\mathfrak{p})}$ to the different completions. For $x \in (\mathbb{L}_n)_p$ they can be interpreted as components of the vector residues under the Chinese Remainder Theorem, so we write $x_i = \iota_i(x) \in (\mathbb{L}_n)_{\nu^i \mathfrak{p}}$ and $x = (x_0, x_1, \ldots, x_{p-1})$. We have $(\mathbb{L}_n)_{\nu^i \mathfrak{p}} \cong (\mathbb{K}_n)_p$ and the action of ν on the vector representation of x is $\nu(x) = (x_{p-1}, x_0, \ldots, x_{p-2})$, thus a cyclic right shift. As a consequence, we obtain the following representation of the norm:

$$\mathcal{N}(x) = \mathcal{N}(x_0, x_1, \ldots, x_{p-1}) = \prod_{i=0}^{p-1} x_i.$$

In particular, for the diagonal embedding of $x \in \mathbb{L}_n$ we have $\iota_i(x) \equiv \nu^{-i}(x) \bmod \mathfrak{p}_n$. The decomposition of $U(\mathbb{K}_n)_p$ according to the orthogonal idempotents induces an important decomposition of $U(\mathbb{L}_n)$, namely,

$$U_k(\mathbb{L}_n) = \\ \left\{x = (x_0, x_1, \ldots, x_{p-1}) \in U(\mathbb{L}_n) \ : \ x_i \in U(\mathbb{K}_n)^{e_{2k}} \right\}. \quad (11)$$

It follows from remarks above that $U_k(\mathbb{L}_n) \subset U'(\mathbb{L}_n)$ for all $k > 0$. We define accordingly a sub-decomposition of $U_0(\mathbb{L}_n)$ as follows:

$$\begin{aligned} \overline{S} &= \{x \in U(\mathbb{L}_n) \ : \ x_i \in \mathbb{Z}_p\} \cong (\mathbb{Z}_p)^p, \\ \tilde{U}_0(\mathbb{L}_n) &= \{x \in U(\mathbb{L}_n) \ : \ x_i \in \Delta_n^{\Lambda} \}, \\ \Omega(\mathbb{L}_n) &= \tilde{U}_0(\mathbb{L}_n) \bigoplus \left(\oplus_{k=1}^{d-1} U_k(\mathbb{L}_n)\right). \end{aligned}$$

We define in addition a list of special elements of $U(\mathbb{L}_n)$, written in their vector representation, which act as generators of the $\mathbb{Z}_p[s]$-submodules above

$$\begin{aligned} \Xi_n^{(k)} &= (\xi_{n,2k}, 1, \ldots, 1), \quad k > 1, \quad \Xi_n^{(0)} = (\upsilon, 1, \ldots, 1), \\ \Theta_n &= (\Delta_n, 1, \ldots, 1). \end{aligned}$$

With these notation, the structure of the units $U(\mathbb{L}_n)$ is given by the following:

Lemma 2 *The following direct sum decompositions*

$$U(\mathbb{L}_n) = \bigoplus_{i=0}^{d}, \quad U_k(\mathbb{L}_n) = \overline{S} \bigoplus \Omega_n, \quad (12)$$

and the following generation relations hold

$$\begin{aligned} \overline{S} &= (\Xi_n^{(0)})^{\mathbb{Z}_p[s]}, \quad \tilde{U}_0(\mathbb{L}_n) = \Theta_n^{\Lambda[s]}, \\ U_k(\mathbb{L}_n) &= (\Xi_n^{(k)})^{\Lambda[s]}, k > 0 \quad forall\, n \geq 0. \end{aligned} \quad (13)$$

Proof The relations (13) are verified directly from the definition, using the circular shift action of ν on the vector representation of elements in $U(\mathbb{L}_n)$. The orthogonality of the idempotents induces a direct sum decomposition $U(\mathbb{L}_n) = \bigoplus_{k=0}^{d} U_k(\mathbb{L}_n)$, and $U_0(\mathbb{L}_n) = \overline{S} \oplus \tilde{U}_0(\mathbb{L}_n)$ follows from Lemma 1. This induces the decomposition $U(\mathbb{L}_n) = \overline{S} \oplus \Omega_n$, and completes the proof. □

Note that $\Omega_n \subsetneq U_n'$. In view of the above decomposition, we denote the projections $\overline{\iota}_s : U(\mathbb{L}_n) \to \overline{S}$ and $\overline{\iota}_r : U(\mathbb{L}_n) \to \Omega_n$, the singular and regular projections of $U(\mathbb{L}_n)$. The modules are accordingly the singular, respective the regular module of $U(\mathbb{L}_n)$. We note that the norms $\mathbf{N}_{n,0} : \Omega_n \to \Omega_0$ are surjective, while $\cap_n \mathbf{N}_{n,0}(U(\mathbb{L}_n)) = \Omega_0$, so the singular module is the natural norm defect of the universal norms of local units from $U(\mathbb{L}_n)$. Moreover, $\mathcal{N}(\Omega_0) = U'(\mathbb{K}_n) \supset \mathcal{N}(\overline{E}(\mathbb{L}))$. This motivates the denomination.

The notion of *singularity level* defined in the following applies to global units:

Definition 4 For $x \in U'(\mathbb{L}_n)$, we define the integer $\ell(x)$ as the largest integer l for which $\iota_s(x) \in \Xi_0^{s^l \mathbb{Z}_p[s]}$. If $C \subset U'(\mathbb{L}_n)$ is some submodule, $\ell(C) = \min_{x \in C}(\ell(x))$ and $\ell(\mathbb{L}_n) := \ell(E(\mathbb{L}_n))$. A unit $e \in E(\mathbb{L}_n)$ is called *singular*, if $\iota_s(e) \in \overline{S}$ generates $\iota_s(\overline{E}(\mathbb{L}_n))$ as a $\mathbb{Z}_p[s]$-module; in particular, $\ell(e) = e(\mathbb{L}_n)$ is minimal over all units in $E(\mathbb{L}_n)$

In the sequel, we gather some useful facts about the local algebras:

Lemma 3 *The following facts hold in the algebras* $(\mathbb{L}_n)_p$*:*

A. *In* $(\mathbb{L}_n)_p$*, we have*

$$\begin{aligned} Ker(s : (\mathbb{L}_n)_p \to (\mathbb{L}_n)_p) &= (\mathbb{K}_n)_p, \text{ and} \\ Ker(\mathcal{N} : (\mathbb{L}_n)_p \to (\mathbb{K}_n)_p) &= (\mathbb{L}_{n}{}_p^{\times})^s. \end{aligned}$$

B. *The following* identity principle *holds: if* $\alpha, \beta \in \mathbb{L}_n$ *are such that* $\iota_k(\alpha) = \iota_k(\beta)$ *for some* $k \in \{0, 1, \ldots, d\}$*, then* $\alpha = \beta$*. In particular, if* $\alpha \notin \mathbb{Q}$*, then* $\iota_k(\alpha) \notin \mathbb{Q}$*, and for* $\varepsilon \in E(\mathbb{L})$ *we have* $\iota_k(\varepsilon) = \pm 1$ *iff* $\varepsilon = \pm 1$*.*

C. *For* $x, y \in U'(\mathbb{L}_n)$ *and* $m \geq 0$*, we have* $\ell(x^{s^m}) = \ell(y^{s^m})$ *iff* $\ell(x) = \ell(y)$*. Also,* $\ell(E(\mathbb{L}_n)) \geq \ell(E(\mathbb{L}_{n'}))$ *for* $n > n'$*.*

D. *If* $x \in \mathbb{L}_n$ *has* $\ell(x) = 0$*, then* $x \notin E(\mathbb{L}_n)$*.*

Proof If $x \in \mathrm{Ker}(s : (\mathbb{L}_n)_p \to (\mathbb{L}_n)_p)$, then $\nu(x) = x$ and thus all the components of x under the Chinese Remainder Theorem are identical. There is thus some $y \in (\mathbb{K}_n)_p$ such that $x = (y, y, \ldots, y)$, which is the diagonal representation of the lift of y to $\mathbb{L}_n$; this confirms the first claim in A. For the second, let $x = (x_0, x_1, \ldots, x_{p-1}) \in \mathrm{Ker}(\mathcal{N} : (\mathbb{L}_n)_p \to (\mathbb{K}_n)_p)$. Then $\prod_i x_i = 1$ and thus $1/x_{p-1} = \prod_{i=0}^{p-2} x_i$. We define $y \in (\mathbb{L}_n)_p$ such that $y^s = x$ as follows. Let $y_0 = t$ and define inductively: $y_1 = x_0 y_0 = x_0 t; \quad y_2 = x_1 y_1, \quad \ldots, \quad y_{p-1} = x_{p-2} \cdot y_{p-2}$. We notice that $y_k = t \cdot \prod_{l=0}^{k-1} x_i$ for $k = 1, 2, \ldots, p-1$ and the conditions are consistent with $y_0 = x_{p-1} y_{p-1}$. Let thus $t = 1$ and y be given by $\iota_{k+1}(y) = \prod_{i=0}^{k} x_i$ for $k = 0, 1, \ldots, p-1$, were $\iota_p = \iota_0$. Then $y^s = x$, which confirms the second claim in point A.

For point B, we can assume without loss of generality that $\alpha, \beta \in \mathcal{O}(\mathbb{L}_n)$, since multiplication by a rational integer leaves the identity between components unchanged. Assume thus that $\iota_k(\alpha - \beta) = 0$ for some $k \geq 0$; this implies that $\alpha - \beta \equiv 0 \bmod \mathfrak{P}_n^N$ for all $N > 0$. Consequently $\mathbf{N}_{\mathbb{L}_n/\mathbb{Q}}(\alpha - \beta) \equiv 0 \bmod p^N$ for all N, which implies $\alpha - \beta = 0$, as claimed. The further special cases in the claim B. are direct consequences of this fact.

For point C., let $x_s = \iota_s(x), y_s = \iota_s(y) \in \Xi_0^{s\mathbb{Z}_p[s]}$. Assume that $\ell(x^{s^m}) = \ell(y^{x^m})$. Then there are units $V, V_1 \in (\mathbb{Z}_p[s])^\times$ and a $l > m$ such that $x_s^{s^m} = y_s^{s^m V} = \Xi_0^{s^l V_1}$ and $(\Xi_0^{s^{l-m} V_1}/x_s)^{s^m} = 1$. But $\Xi_0^{s\mathbb{Z}_p[s]}$ is $\mathbb{Z}_p$-free, so it follows that $x_s = \Xi_0^{s^{l-m} V_1}$; likewise, $y_s = \Xi_0^{s^{l-m} V_1/V}$, and it follows $\ell(x) = \ell(y)$. The other direction is straightforward. Moreover, there are some units $e \in E(\mathbb{L}_n)$, $e' \in E(\mathbb{L}'_n)$ such that $\ell(e) = \ell(E(\mathbb{L}_n))$ and $\ell(e') = \ell(E(\mathbb{L}_{n'}))$. For $n > n'$, obviously $e' \in E(\mathbb{L}_n)$ and thus $\ell(\mathbb{L}_{n'}) = \ell(e) \leq \ell(\mathbb{L}_n)$, which completes the proof of point C.

Finally assume that $\ell(x) = 0$, so $x = \Xi_0^V \cdot \omega$, with $V \in (\mathbb{Z}_p[s])^\times$ and $\omega \in \Omega_n$. Then $\mathbf{N}_{\mathbb{L}_n/\mathbb{K}}(x) = (1+p)^{V(0)p^n} \cdot \omega'$, where $\omega' \in \Omega_0 \cap U(\mathbb{K}) \subset U'(\mathbb{K})$. Consequently, $\mathbf{N}_{\mathbb{L}_n/\mathbb{Q}}(x) = (1+p)^{V(0)(p-1)p^n}$, and this norm is only a unit in $\mathbb{Z}$, if $V(0) = 0$, which is inconsistent with V being a unit. This completes the proof of point D. □

3 Existence of a Singular Capitulation Unit

The purpose of this chapter is to prove the

Proposition 1 *Under the above assumptions, for all sufficiently large n, there are singular capitulation units* $\delta_n \in E(\mathbb{L}_n)$, *verifying* $\ell(\delta_n) = \ell(\mathbb{L}_n)$ *and* $\mathcal{N}(\delta_n) = 1$, *while* $\delta_n \notin E(\mathbb{L}_n)^{(s,p)}$.

We first note the following

Lemma 4 *Assuming that* $|A(\mathbb{K}_n)| = |A(\mathbb{K}_m)|$ *for all* $n \geq m$, *we have*

$$E(\mathbb{K}_n) = C(\mathbb{K}_n) \cdot E(\mathbb{K}_m), \quad \forall n \geq m. \tag{14}$$

Moreover, there are uniformizors $\pi_n \in E'(\mathbb{L}_n)$ *with* $\mathcal{N}(\pi_n) = \wp_n$ *and consequently* $C(\mathbb{K}_n) \subset \mathcal{N}(E(\mathbb{L}_n))$.

Proof By the analytic class number formula, the hypothesis of this lemma implies that the index $|(E(\mathbb{K}_n)/C(\mathbb{K}_n))_p|$ is stable for all $n \geq m$, being equal to the stabilized size of the p-part of the class group. Since the cyclotomic units are norm coherent, the claim (14) follows.

We choose now $k > 0$ sufficiently large, such that $E(\mathbb{K}_m)^{p^k} \subset C(\mathbb{K}_m)$ and let $\rho_k \in E'(\mathbb{L}_{n+k})$ have norm $\mathcal{N}(\rho') = \lambda_{n+k} e$, with $e \in E(\mathbb{K}_{n+k})$. The choice of k implies that $\mathbf{N}_{n+k,n}(e) \in C(\mathbb{K}_n)$ and consequently, the p-unit $\rho := \mathbf{N}_{n+k,n}(\rho')$ verifies $\mathcal{N}(\rho) \in \lambda_n C(\mathbb{K}_n) = \lambda_n^{1+(\sigma-1)\mathbb{Z}[G_n]}$. We may define thus $\pi_n = \rho^t$ for some adequate $t \in \mathbb{Z}[G_n]$, which implies the claim. □

We proceed with the proof of the Proposition 1:

Proof Let $n > m$ and $d_n \in E(\mathbb{L}_n)$ have $\ell(d_n) = \ell(\mathbb{L}_n) = L$. Let $D_n = \mathcal{N}(d_n) = \varepsilon \cdot \gamma_n$, with $\varepsilon \in E(\mathbb{K}_m)$ and $\gamma_n \in C(\mathbb{K}_n)$, a decomposition based on (14). Let now $N > n$ be sufficiently large and choose $\gamma_N \in C(\mathbb{K}_N)$ such that $\mathbf{N}_{N,n}(\gamma_N) = \gamma_n$.

We deduce from Lemma 4, that there is some $\gamma'_N \in E(\mathbb{L}_N)$, with $\gamma_N = \mathcal{N}(\gamma'_N)$. Let $\gamma'_n = \mathbf{N}_{N,n}(\gamma'_N)$. We may choose N sufficiently large, so that $\ell(\gamma'_n) = 2L$, say. Since $\mathcal{N}(\gamma'_n) = \gamma_n$, by construction, by letting $d'_n = d_n/\gamma'_n$, we obtained a unit with $\ell(d'_n) = \ell(d_n)$ and $\mathcal{N}(d'_n) = \varepsilon_m$. Since $(E_m/C_m)_p^{e_0} = 1$, while the singular module $S \subset \overline{E}(\mathbb{K}_n)^{e_0}$, we may let $\delta'_n = (d'_n)^{e'_0}$ for an approximation $e'_0 \in \mathbb{Z}[G]$ with $e'_0 \equiv e_0 \bmod p^M \mathbb{Z}_p[G]$. Then $D'_n := \mathcal{N}(\delta'_n) \in U(\mathbb{K}_n)^{p^M}$ while $\ell(\delta'_n) = \ell(d_n)$. By taking M sufficiently large, we can reach $D'_n = v^p$, for some unit $v \in E(\mathbb{K}_n)$. Finally, letting $\delta_n = \delta'_n/v$ we obtained a unit with $\mathcal{N}(\delta_n) = 1$ and $\ell(\delta_n) = \ell(d_n)$. The minimality of the index $\ell(d_n)$ implies that $\delta_n \notin E(\mathbb{L}_n)^{(p,s)}$. Thus δ_n satisfies all the claims of Proposition 1, and this completes the proof. □

Finally, we combine the previous result with the Theorem Hilbert 94, thus obtaining

Corollary 1 *There is a* singular class $a_n \in A(\mathbb{K}_n)[p]$, $a_n \neq 1$, *such that the primes of a_n become principal in $\mathbb{L}_n$ and for each $\mathfrak{P} \in a_n$, if $\mathfrak{P}\mathcal{O}(\mathbb{L}_n) = (\rho)$, then $\rho^s \in \delta_n \cdot E(\mathbb{L}_n)^s$.*

Proof The properties of δ_n imply, by Hilbert's Theorem 94, that there is an ideal $\mathfrak{B} \subset \mathcal{O}(\mathbb{K}_n)$, which is not principal in $\mathbb{K}_n$ but capitulates in $\mathbb{L}_n$ – say, $\mathfrak{B} \cdot \mathcal{O}(\mathbb{L}_n) = (\beta)$ – such that $\beta^s = \delta_n$. Let then $a = [\mathfrak{B}]$ be the class of this ideal and let $\mathfrak{P} \in a$, so there is some $\alpha \in K_n$ such that $\mathfrak{P} = (\alpha)\mathfrak{B}$. Then $(\rho) = (\alpha) \cdot (\beta)$ and there is a unit $d \in E(\mathbb{L}_n)$, such that $\rho = d\alpha \cdot \beta$. Since $\alpha \in \mathbb{K}_n$, we conclude that $\rho^s = d^s \cdot \beta^s = d^s \cdot \delta_n \in \delta_n E(\mathbb{L}_n)^s$, as claimed. □

4 Proof of the Main Theorem

The assumption that $A(\mathbb{K}_\infty)$ is bounded implies that there is an n_0 such that for all $n > n_0$, we have $A(\mathbb{K}_n) \cong A(\mathbb{K}_{n+1})$ and thus

$$\begin{aligned} p - \text{rank } \left(\text{Ker} : \iota_{n,n+1} A(\mathbb{K}_n) \to A(\mathbb{K}_{n+1})\right) &= p - \text{rank } (A(\mathbb{K}_n)) \\ &= p - \text{rank } (A(\mathbb{K}_{n+1})). \end{aligned}$$

Furthermore, Corollary 1 implies that

Lemma 5 *For $n > n_0$, the units δ_n are singular and at the next level, $\delta_n \in E(\mathbb{L}_{n+1})^s$.*

Proof For $n > n_0$, we let $\mathfrak{A}_n \in a_n$, where a_n is the class defined in Corollary 1, and assume that $\delta_n = \gamma_n^s$, with $(\gamma_n) = \mathcal{O}(\mathbb{L}_n) \cdot \mathfrak{A}$. Since $n > n_0$ and $a_n^p = 1$, the ideal $\mathfrak{A}_n$ capitulates in $\mathbb{K}_{n+1}$. There is thus an $\alpha_{n+1} \in \mathbb{K}_{n+1}$ such that $(\alpha_{n+1}) = \mathfrak{A}_n \cdot \mathcal{O}(\mathbb{K}_{n+1})$. But then $\alpha_{n+1}\mathcal{O}(\mathbb{L}_{n+1}) = \gamma_n \mathcal{O}(\mathbb{L}_{n+1})$ and there is a unit $\varepsilon_{n+1} \in E(\mathbb{L}_{n+1})$ such that $\gamma_n = \varepsilon \cdot \alpha_{n+1}$. Then $\delta_n = \gamma_n^s = \varepsilon^s$ in $E(\mathbb{L}_{n+1})^s$, which completes the proof. □

We now give the proof of Theorem 1:

Proof Let $n > n_0$. Since δ_n is singular, we have $\ell(\mathbb{L}_n) = \ell(\delta_n)$. On the other hand, $\delta_n = \varepsilon^s_{n+1}$, as shown in Lemma 5 above, and point 3 of Lemma 3 implies herewith that

$$\ell(\mathbb{L}_{n+1}) \leq \ell(\varepsilon_{n+1}) < \ell(\delta_n) = \ell(\mathbb{L}_n).$$

For large enough n, the vertical capitulation has thus as effect a decrease of $\ell(\mathbb{L}_n)$. Since $\ell(\mathbb{L}_{n_0})$ is an integer, we reach in finitely many steps a contradiction; there is indeed some $n_1 \leq n_0 + \ell(\mathbb{L}_{n_0})$ such that $\ell(\mathbb{L}_{n_1}) = 0$, a contradiction to point D in Lemma 3. This completes the proof of Theorem 1. □

References

1. G. Banaszak, W. Gajda, On the arithmetic of cyclotomic fields and the K-theory of $\mathbf{Q}$, in *Algebraic K-Theory*, volume 199 of Contemporary Mathematics; Am. Math. Soc. (1996)
2. D. Hilbert, *The Theory of Algebraic Numbers (Zahlbericht)* (Springer, 1998)
3. M. Kurihara, Some remarks on conjectures about cyclotomic fiels and k-groups of $\mathbb{Z}$. Compos. Math. **81**, 223–236 (1992)
4. S. Lang. *Cyclotomic Fields I and II*, volume 121 of Graduate Texts in Mathematics. Springer, combined Second Edition edition (1990)
5. S. Sitaraman, Vandiver revisited. J. Number Theory **57**(1), 122–129 (1996)
6. C. Soulé, Perfect forms and Vandiver's conjecture. J. Reine und Angew. Math. **517**, 209–221 (1999)
7. L. Washington. *Introduction to Cyclotomic Fields*, volume 83 of Graduate Texts in Mathematics. Springer (1996)

Heights and Principal Ideals of Certain Cyclotomic Fields

René Schoof

1 Introduction

Any prime number l splits completely in the cyclotomic field $\mathbf{Q}(\zeta_{l-1})$. The primes lying over l all have norm l and are Galois conjugate. Consider the following set of prime numbers:

$$S = \{2, 3, 5, 7, 11, 13, 17, 19, 23, 29, 31, 37, 41, 43, 61, 67, 71\}.$$

In this expository note we give a self-contained proof of the following theorem

Theorem 1.1 *For a prime number l the following are equivalent.*

(i) $l \in S$;
(ii) the class number of $\mathbf{Q}(\zeta_{l-1})$ *is* 1;
(iii) The prime ideals lying over l in $\mathbf{Q}(\zeta_{l-1})$ *are principal.*

It is trivial that (ii) implies (iii). The fact that (i) implies (ii) is not trivial, but it is standard. In fact, using Odlyzko's [5] discriminant bounds, Masley and Montgomery [4] determined in the 1970's all cyclotomic fields with class number 1. See [7]. For proving that (i) implies (ii) one needs much less. We work this out in Sect. 3.

A proof of the fact that (iii) implies (i) was recently published by Bernat Plans [6]. It is an application of a theorem, proved in 2000 by Amoroso and Dvornicich [1], supplemented by computations by Hoshi [2]. In their paper, Amoroso and Dvornicich themselves already had used their theorem in a similar way proving that certain cyclcotomic fields have nontrivial class numbers. We prove a weak version of their theorem in Sect. 2.

Condition (iii) of Theorem 1.1 first came up in a 1974 paper by Lenstra [3] on a problem related to Noether's problem and the inverse problem of Galois theory.

R. Schoof (✉)
Dipartimento di Matematica, Università di Roma Tor Vergata, I-00133 Roma, Italy
e-mail: schoof.rene@gmail.com

K. Chakraborty et al. (eds.), *Class Groups of Number Fields and Related Topics*,
https://doi.org/10.1007/978-981-15-1514-9_8

Lenstra showed that the set of prime numbers satisfying the condition has Dirichlet density zero [3, , Cor.6.7].

We deduce Theorem 1.1 in Sect. 4 from the results in Sects. 2 and 3.

This note is based on an expository lecture given at the ICCGNFRT meeting at the HRI, Allahabad, September 2017.

2 Heights

We recall some basic properties of heights. For every finite or infinite prime v of a number field F, let $|x|_v$ denote the corresponding normalized valuation of $x \in F^*$. This means that for finite primes v we put $|x|_v = q^{-v(x)}$, where q is the cardinality of the residue field. For infinite real primes we use the usual absolute value and for complex primes its square.

Then the *product formula* holds: for every $x \in F^*$ we have

$$\prod_v |x|_v = 1.$$

For any positive real t we put $\log^+ t = \max(\log t, 0)$. The *height* $h(x)$ of $x \in F^*$ is defined as

$$h(x) = \sum_v \log^+ |x|_v.$$

Note that the value of $h(x)$ depends not only on x but also on the number field F. The *absolute height*

$$\frac{h(x)}{[F:\mathbf{Q}]}$$

is independent of F and depends only on x.

It is easy to see that for all $x, y \in F^*$ and every prime v we have

$$|x-y|_v \le 2^{u_v} \max(1, |x|_v) \cdot \max(1, |y|_v),$$

where $u_v = 0$, 1 or 2, depending on whether v is finite, real or complex, respectively. Indeed, by symmetry we may assume that $|x|_v \ge |y|_v$. Then the triangle inequality implies that $|1 - y/x|_v$ is at most 2^{u_v}. It follows that $|x-y|_v \le 2^{u_v}|x|_v$ and the inequality follows.

Sharper upper bounds for $|x-y|_v$ give rise to lower bounds for the heights of either x or y.

Proposition 2.1 *Let F be a number field and let x and y be distinct elements of F^*. For every prime v, let $0 < c_v \le 1$. If*

$$|x-y|_v \le 2^{u_v} c_v \cdot \max(1, |x|_v) \cdot \max(1, |y|_v), \qquad \textit{for all primes } v.$$

Then

$$h(x)+h(y)\geq -[F:\mathbf{Q}]\log 2-\sum_{v}\log c_v.$$

Proof By the product formula and the inequalities of the hypothesis we have

$$0=\sum_{v}\log|x-y|_v\leq\sum_{v}\log(2^{u_v}c_v)+h(x)+h(y).$$

The result then follows from the fact that $\sum_v u_v=\sum_{v\text{ infinite}}u_v=[F:\mathbf{Q}]$.

The following lemma is used in the proof of the result by Amoroso and Dvornicich.

Lemma 2.2 *Let F be a number field, let v be a finite prime of F and let $\chi,\chi':F^*\longrightarrow F^*$ be two homomorphisms that preserve v-integrality. Let $c\in\mathbf{R}_{>0}$. If we have*

$$|\chi(\alpha)-\chi'(\alpha)|_v\leq c,\qquad \textit{for all non-zero }\alpha\in O_F,$$

then

$$|\chi(\alpha)-\chi'(\alpha)|_v\leq c\cdot\max(1,|\chi(\alpha)_v)\cdot\max(1,|\chi'(\alpha)|_v),\qquad \textit{for all }\alpha\in F^*.$$

Proof Let $\alpha\in F^*$. By the Chinese remainder theorem, we can find an element $\beta\in O_F$ for which $\alpha\beta\in O_F$ and $|\beta|_v=\max(1,|\alpha|_v)^{-1}$. Since χ preserves v-integrality, this implies that $|\chi(\beta)|_v=\max(1,|\chi(\alpha)|_v)^{-1}$. From the identity

$$\chi(\alpha)-\chi'(\alpha)=\frac{1}{\chi(\beta)}\left(\chi(\alpha\beta)-\chi'(\alpha\beta)+\chi'(\alpha)\chi'(\beta)-\chi'(\alpha)\chi(\beta)\right),$$

we deduce the inequality

$$|\chi(\alpha)-\chi'(\alpha)|_v\leq\frac{c}{|\chi(\beta)|_v}\max(1,|\chi'(\alpha)|_v)=c\max(1,|\chi(\alpha)|_v)\max(1,|\chi'(\alpha)|_v),$$

as required.

Proposition 2.3 *(Amoroso and Dvornicich [1]) Let m be a positive integer and let ζ_m denote a primitive m-th root of unity. Suppose that $\alpha\in\mathbf{Q}(\zeta_m)^*$ is not a root of unity. Then for every prime number p we have*

$$\frac{h(\alpha)}{[F:\mathbf{Q}]}\geq\frac{\log(p/2)}{2p}.$$

If p does not divide m, we have the sharper estimate

$$\frac{h(\alpha)}{[F:\mathbf{Q}]}\geq\frac{\log(p/2)}{p+1}.$$

Proof Put $F = \mathbf{Q}(\zeta_m)$. If p does not divide m, we apply Proposition 2.1 to $x = \alpha^p$, $y = \sigma(\alpha)$ and $c_v = |p|_v$ when v lies over p, while $c_v = 1$ for the other primes v. Here σ is the Frobenius automorphism in $\mathrm{Gal}(F/\mathbf{Q})$ of the primes lying over p. It fixes every v lying over p. Since $h(\alpha^p) = ph(\alpha)$ and $h(\sigma(\alpha)) = h(\alpha)$, the second estimate then follows.

It remains to check that $x = \alpha^p$, $y = \sigma(\alpha)$ satisfy the hypotheses of Proposition 2.1. Since α is not a root of unity, the elements x and y are distinct. In order to check the inequality in the condition of Proposition 2.1, we recall that the ring of integers of F is $\mathbf{Z}[\zeta_m]$. The fact that $\sigma(\zeta_m) = \zeta_m^p$, implies therefore that $\sigma(\alpha) \equiv \alpha^p \pmod{p}$ for all integral α. This implies that the inequality holds for integral $x = \sigma(\alpha)$ and $y = \alpha^p$. An application of Lemma 2.2 to the homomorphisms $\chi(\alpha) = \sigma(\alpha)$ and $\chi'(\alpha) = \alpha^p$ shows that it also holds for all $\alpha \in F^*$ and we are done.

If p divides m, we we apply Proposition 2.1 to $x = \alpha^p$, $y = \sigma(\alpha)^p$ and and $c_v = |p|_v$ when v lies over p, while $c_v = 1$ for the other primes v. Here σ generates the Galois group of F over its subfield $\mathbf{Q}(\zeta_{m/p})$. The first inequality follows readily.

It remains to check the hypotheses of Proposition 2.1. Since σ fixes $\mathbf{Q}(\zeta_{m/p})$, we have $\sigma(\zeta_m) = \zeta_m^t$ for some $t \equiv 1 \pmod{m/p}$. It follows that $\sigma(\zeta_m)^p = \zeta_m^p$ and hence $\sigma(\alpha)^p \equiv \alpha^p \pmod{p}$ for all $\alpha \in \mathbf{Z}[\zeta_m]$. In other words, the inequality in the hypothesis of Proposition 2.1 holds for $x = \sigma(\alpha)^p$ and $y = \alpha^p$ for every integral $\alpha \in F$. An application of Lemma 2.2 to the homomorphisms $\chi(\alpha) = \sigma(\alpha)^p$ and $\chi'(\alpha) = \alpha^p$ shows that the inequality holds for all $\alpha \in F^*$.

Finally, if x and y were equal, then $\alpha = \sigma(\alpha)\zeta'$ for some $\zeta' \in \mu_p$. The kernel of the homomorphism $\mu_m \longrightarrow \mu_m$ given by $\xi \mapsto \sigma(\xi)/\xi = \xi^{t-1}$, is $\mu_{m/p}$. Therefore the image is μ_p. It follows that $\zeta' = \sigma(\xi)/\xi$ for some $\xi \in \mu_m$. This means that $\xi\alpha$ is fixed by σ and is hence contained in the subfield $\mathbf{Q}(\zeta_{m/p})$. Since α and $\xi\alpha$ have the same height, we may replace α by $\xi\alpha$ and $F = \mathbf{Q}(\zeta_m)$ by $\mathbf{Q}(\zeta_{m/p})$. We repeat this until either $x \neq y$, in which case all conditions of Proposition 1 are satisfied, or until p does not divide m, in which case we have the sharper estimate that we already proved.

Corollary 2.4 *Let l be a prime number and suppose that the prime ideals of $\mathbf{Q}(\zeta_{l-1})$ lying over l are principal. Then we have*

$$\frac{\log l}{\phi(\ell - 1)} \geq \frac{\log(5/2)}{10},$$

where ϕ is Euler's function. Moreover, for any prime p for which $l \not\equiv 1 \pmod{p}$, we have

$$\frac{\log l}{\phi(\ell - 1)} \geq \frac{\log(p/2)}{p+1}.$$

Proof We put $F = \mathbf{Q}(\zeta_{l-1})$ and, as in [1, Cor.1], we put $\alpha = \overline{\pi}/\pi$, where π is a generator of a prime of F lying over l. Since l splits completely in F, the quotient $\overline{\pi}/\pi = \alpha$ is not a root of unity. Since $h(\alpha) = \log l$, an application of Proposition 2.3 implies the result.

Remark 2.5 For $p = 2$, the bounds of Proposition 2.3 are trivial. However, one can obtain nontrivial bounds by observing that for $\alpha \in \mathbf{Z}[\zeta_m]$ one has $\sigma(\alpha)^2 \equiv \alpha^4 \pmod 4$ when $m \not\equiv 0 \pmod 4$ and σ is the Frobenius automorphism of the primes lying over 2. When $m \equiv 0 \pmod 4$ and σ is the automorphism of $\mathbf{Q}(\zeta_m)$ for which $\sigma(\zeta_m) = \zeta_m^{1+m/2} = -\zeta_m$, one has $\sigma(\alpha)^2 \equiv \alpha^2 \pmod 4$. This leads to the inequality

$$\frac{h(\alpha)}{[F:\mathbf{Q}]} \geq \frac{\log(2)}{6},$$

for all m and all $\alpha \in \mathbf{Q}(\zeta_m)^*$ that are not a root of unity.

Remark 2.6 In the proof of Proposition 2.3 of the case where p divides m, one may actually take $c_v = |p|_v^{p/(p-1)}$ for the primes v lying over p. This is slightly smaller and gives a better estimate in Corollary 2.4. It makes little difference for the proof of Theorem 1.1.

3 Discriminant Bounds

In this section, we explain how to prove the implication (i) $\Rightarrow$ (ii) of the main theorem. We use Odlyzko's discriminant bounds [5].

In general, the class number of a cyclotomic field $\mathbf{Q}(\zeta_m)$ is the product of the class number of the maximal real subfield $\mathbf{Q}(\zeta_m)^+$ of $\mathbf{Q}(\zeta_m)$ and the so-called *relative class number*. The latter is a product of generalized Bernoulli numers and is easy to compute [7, Theorem 4.17]. It is an easy matter to check that for the primes in the set S of Theorem 1.1, the relative class numbers of $\mathbf{Q}(\zeta_{l-1})$ are all equal to 1. This is left to the reader, who may prefer to consult the table in [7, p.412]. To show that the class numbers themselves are also 1, it suffices to show that the class numbers of the subfields $\mathbf{Q}(\zeta_m)^+$ are 1.

The absolute degree of $\mathbf{Q}(\zeta_m)$ over $\mathbf{Q}$ is $\phi(m)$. The root discriminant δ_m of $\mathbf{Q}(\zeta_m)$ is the $\phi(m)$-th root of the absolute value of its discriminant. Explicitly, δ_m is equal to $m \prod_p p^{-1/(p-1)}$, where the product runs over the prime divisors of m. See [7, Proposition 2.7]. For $m > 2$, the subfield $\mathbf{Q}(\zeta_m)^+$ has absolute degree $\frac{1}{2}\phi(m)$, while its root discriminant is at most δ_m.

Consider the set S of primes of Theorem 1.1. For the primes $l = 2, 3, 5, 7, 11$ and 13, the field $\mathbf{Q}(\zeta_{l-1})^+$ is either $\mathbf{Q}$ or one of the quadratic fields $\mathbf{Q}(\sqrt{3})$ or $\mathbf{Q}(\sqrt{5})$. It is well known and easy to verify that the class numbers of these fields are equal to 1. This leaves us with the primes $l = 17, 19, 23, 29, 31, 37, 41, 43, 61, 67$ and 71.

In Table 1 we list the degrees and root discriminants of these fields.

The root discriminant of any totally real number of degree d is bounded below by Odlyzko's discriminant bound Odl(d). See [7, , 11.4]. The function Odl((d) is monotonically increasing. For degree $d \leq 14$, we list its values, or rather approximations to them, in Table 2. See also [5].

Table 1 Degrees and root discriminants of $\mathbf{Q}(\zeta_{l-1})$

l	$\phi(l-1)$	δ_{l-1}		l	$\phi(l-1)$	δ_{l-1}
17	8	8.000		41	16	13.375
19	6	5.197		43	12	8.767
23	10	8.655		61	16	11.583
29	12	10.123		67	20	14.991
31	8	5.792		71	24	16.923
37	12	10.393				

Table 2 Odlyzko's bounds

d	Odl(d)	d	Odl(d)	d	Odl(d)	d	Odl(d)
1	0.996	5	6.514	9	11.787	13	16.044
2	2.222	6	7.926	10	12.941	14	16.971
3	3.609	7	9.279	11	14.034		
4	5.062	8	10.568	12	15.068		

The Hilbert class field of $\mathbf{Q}(\zeta_{l-1})^+$ is totally real. Its degree over $\mathbf{Q}(\zeta_{l-1})^+$ is equal to the class number of $\mathbf{Q}(\zeta_{l-1})^+$. Since it is an everywhere unramified extension of $\mathbf{Q}(\zeta_{l-1})^+$, its root discriminant is equal to the root discriminant of $\mathbf{Q}(\zeta_{l-1})^+$, which is at most δ_{l-1}. Therefore, we can use Odlyzko's bounds to bound the class number h of $\mathbf{Q}(\zeta_{l-1})^+$. To be precise, we have

$$h\phi(l-1)/2 < d,$$

for any d for which $\mathrm{Odl}(d)$ exceeds δ_{l-1}. It follows easily from the entries in the two tables that $h < 2$ in each case. For instance, for $l = 71$, we have $\delta_{l-1} = 16.923\ldots$. Since $\mathrm{Odl}(14) = 16.971$, we may take $d = 14$ and we find that $h \cdot \frac{1}{2} \cdot 24 < 14$.

This implies that for the primes in the set S of Theorem 1.1, the class numbers of $\mathbf{Q}(\zeta_{l-1})^+$ are equal to 1, as required.

4 Plans' Theorem

In this section, we prove the implication (iii) $\Rightarrow$ (i) of Theorem 1.1.

The degree $[\mathbf{Q}(\zeta_{l-1} : \mathbf{Q}] = \phi(l-1)$ grows faster than $\log l$. In fact, it is easy to prove that $\phi(l-1) \geq \sqrt{(l-1)/2}$. Therefore the first inequality of Corollary 2.4 can only hold for finitely many primes. It is not difficult to check that the prime numbers l that satisfy the first inequality of Corollary 2.4 are necessarily ≤ 211. An application of the second inequality of Corollary 2.4 with the primes $p \leq 11$ reduces this bound to 79 and excludes $l = 59$. The only primes not in S are $l = 47, 53, 73$ and 79. The relevant cyclotomic fields are $\mathbf{Q}(\zeta_m)$ with $m = 23, 52, 72$ and 39, respectively. We deal with them one by one.

The equation $x^2 + 23y^2 = 4 \cdot 47$ has no solutions in integers. This implies that there is no element of norm 47 in the ring of integers of the quadratic subfield $\mathbf{Q}(\sqrt{-23})$ of $\mathbf{Q}(\zeta_{23})$. This means that the prime ideals over 47 of $\mathbf{Q}(\sqrt{-23})$ are not principal. It follows that the prime ideals over 47 of $\mathbf{Q}(\zeta_{23})$ are not principal either. Similarly, the equation $x^2 + 39y^2 = 4 \cdot 79$ has no solutions in integers. It follows that the prime ideals over 79 of $\mathbf{Q}(\zeta_{39})$ are not principal.

Since the image of the local norm map $\mathbf{Z}_{13}[\zeta_{13}]^* \longrightarrow \mathbf{Z}_{13}^*$ is the group $1 + 13\mathbf{Z}_{13}$, the norm map from $\mathbf{Q}(\zeta_{52})$ to $\mathbf{Q}(i)$ maps numbers that are units at the primes lying over 13 to elements of $\mathbf{Q}(i)^*$ that are congruent to 1 (mod 13). Therefore, the norm map from the class group Cl_{52} of $\mathbf{Q}(\zeta_{52})$ to the (trivial) class group of $\mathbf{Q}(i)$ 'factors' through the ray class group of conductor 13 of $\mathbf{Q}(i)$. In other words, the norm induces a homomorphism

$$N : Cl_{52} \longrightarrow (\mathbf{Z}[i]/(13))^*/\langle i \rangle.$$

It maps the class of an ideal I of $\mathbf{Z}[\zeta_{52}]$ that is prime to 13, to a generator of the ideal $N(I)$ of $\mathbf{Z}[i]$. In particular, any prime of $\mathbf{Z}[\zeta_{52}]$ lying over 53 is mapped to the image of $7 \pm 2i$ in the ray class group. Since $7 \pm 2i$ has order 3 in the group $(\mathbf{Z}[i]/(13))^*/\langle i \rangle$, this image is nontrivial. Therefore the class in Cl_{52} of a prime lying over 53 is not trivial either. It follows that the primes over 53 in $\mathbf{Q}(\zeta_{52})$ are not principal.

Similarly, the image of the local norm map $\mathbf{Z}_3[\zeta_9]^* \longrightarrow \mathbf{Z}_3^*$ is the group $1 + 9\mathbf{Z}_3$. Therefore, the norm map from $\mathbf{Q}(\zeta_{72})$ to $\mathbf{Q}(\sqrt{-2})$ maps numbers that are units at the primes lying over 3 to elements of $\mathbf{Q}(\sqrt{-2})^*$ that are congruent to 1 (mod 9). It follows that the norm maps the class group Cl_{72} of $\mathbf{Q}(\zeta_{72})$ to the ray class group of conductor 9 of $\mathbf{Q}(\sqrt{-2})$. In other words, the norm induces a homomorphism

$$N : Cl_{72} \longrightarrow (\mathbf{Z}[\sqrt{-2}]/(9))^*/\{\pm 1\}.$$

It maps the class of any prime over 73 to the image of $1 \pm 6\sqrt{-2}$ in the ray class group. Since $1 \pm 6\sqrt{-2}$ has order 3 in the group $(\mathbf{Z}[\sqrt{-2}]/(9))^*/\{\pm 1\}$, this image is nontrivial. Therefore the class in Cl_{72} of a prime lying over 73 is not trivial either.

This proves Theorem 1.1.

References

1. F. Amoroso, R. Dvornicich, A lower bound for the height in abelian extensions. J. Number Theory **80**, 260–272 (2000)
2. A. Hoshi, On Noether's problem for cyclic groups of prime order. Proc. Japan Acad. Ser. A **91**, 39–44 (2015)
3. H.W. Lenstra, Rational functions invariant under an abelian group. Invent. Math. **25**, 299–325 (1974)

4. J. Masley, H. Montgomery, Cyclotomic fields with unique factorization. J. Reine Angew. Math. **286**(287), 248–256 (1976)
5. A. Odlyzko, Table 2. Unconditional bounds for discriminants, 29 Nov 1976. http://www.dtc.umn.edu/~odlyzko/unpublished/discr.bound.table2
6. B. Plans, On Noether's rationality problem for cyclic groups over Q. Proc. AMS **145**, 2407–2409 (2016)
7. L.C. Washington, *Introduction to Cyclotomic Fields, Graduate Texts in Math*, vol. 83, 2nd edn. (Springer-Verlag 1997)

Distribution of Residues Modulo p Using the Dirichlet's Class Number Formula

Jaitra Chattopadhyay, Bidisha Roy, Subha Sarkar and R. Thangadurai

1 Introduction

Let p be an odd prime number. A number $a \in \{1, \ldots, p-1\}$ is said to be a *quadratic residue* modulo p, if the congruence

$$x^2 \equiv a \pmod{p}$$

has a solution in $\mathbb{Z}$. Otherwise, a is said to be a *quadratic non-residue* modulo p. The study of distribution of quadratic residues and quadratic non-residues modulo p has been considered with great interest in the literature (see for instance [1, 3–7, 10, 12, 13, 15–25]).

Since $\mathbb{Z}/p\mathbb{Z}$ is a field, the polynomial $X^{p-1} - 1$ has precisely $p-1$ nonzero solutions over $\mathbb{Z}/p\mathbb{Z}$. As p is an odd prime, we see that $X^{p-1} - 1 = (X^{(p-1)/2} + 1)(X^{(p-1)/2} - 1)$ and one can conclude that there are exactly $\frac{p-1}{2}$ quadratic residues as well as non-residues modulo p in the interval $[1, p-1]$.

Question 1 *For an odd prime number p and a given natural number k with $1 \leq k \leq p-1$, we let $S_k = \{a \in \{1, 2, \ldots, p-1\} : a \equiv 0 \pmod{k}\}$ be the subset consisting of all natural numbers which are multiples of k. How many quadratic residues (respectively, non-residues) lie inside S_k?*

J. Chattopadhyay (✉) · B. Roy · S. Sarkar · R. Thangadurai
Harish-Chandra Research Institute, HBNI, Chhatnag Road, Jhunsi 211019, Allahabad, India
e-mail: jaitrachattopadhyay@hri.res.in

B. Roy
e-mail: bidisharoy@hri.res.in

S. Sarkar
e-mail: subhasarkar@hri.res.in

R. Thangadurai
e-mail: thanga@hri.res.in

K. Chakraborty et al. (eds.), *Class Groups of Number Fields and Related Topics*,
https://doi.org/10.1007/978-981-15-1514-9_9

In the literature, there are many papers addressed similar to Question 1 and to name a few, one may refer to [8, 9, 11]. First we shall fix some notations as follows. We denote by $Q(p, S_k)$ (respectively, $N(p, S_k)$) the number of quadratic residues (respectively, quadratic non-residues) modulo p in the subset S_k of the interval $[1, p-1]$.

The standard techniques in analytic number theory answers the above question as

$$Q(p, S_k) = \frac{p-1}{2k} + O(\sqrt{p}\log p) \tag{1}$$

and the same result is true for $N(p, S_k)$ for all k (we shall be proving this fact in this article). However, it might happen that for some primes p, we may have $Q(p, S_k) > N(p, S_k)$ or $Q(p, S_k) < N(p, S_k)$. Using the standard techniques, we could not answer this subtle question. In this article, we shall answer this using the Dirichlet's class number formula for the field $\mathbb{Q}(\sqrt{-p})$, when $k = 2, 3$ or 4. More precisely, we prove the following theorems.

Theorem 1 *Let p be an odd prime. If $p \equiv 3 \pmod 4$, then for any ϵ with $0 < \epsilon < \frac{1}{2}$, we have*

$$Q(p, S_2) - \frac{p-1}{4} \gg_\epsilon p^{\frac{1}{2}-\epsilon}.$$

When the prime $p \equiv 1 \pmod 4$, we have

$$Q(p, S_2) = \frac{p-1}{4}.$$

Corollary 1.1 *Let p be an odd prime and let $\mathcal{O}$ be the set of all odd integers in $[1, p-1]$. If $R = N(p, S_2)$ or $R = Q(p, \mathcal{O})$, then for any ϵ with $0 < \epsilon < \frac{1}{2}$, we have*

$$\frac{p-1}{4} - R \gg_\epsilon p^{\frac{1}{2}-\epsilon}, \text{ if } p \equiv 3 \pmod 4.$$

When the prime $p \equiv 1 \pmod 4$, we have

$$R = \frac{p-1}{4}.$$

Theorem 2 *Let p be an odd prime. If $p \equiv 1, 11 \pmod{12}$, then for any ϵ with $0 < \epsilon < \frac{1}{2}$, we have*

$$Q(p, S_3) - \frac{p-1}{6} \gg_\epsilon p^{\frac{1}{2}-\epsilon}.$$

When $p \equiv 5, 7 \pmod{12}$, in this method, we do not get any finer information other than in (1).

Corollary 1.2 *Let p be an odd prime. If $p \equiv 1, 11 \pmod{12}$, then for any ϵ with $0 < \epsilon < \frac{1}{2}$, we have*

$$\frac{p-1}{6} - N(p, S_3) \gg_\epsilon p^{\frac{1}{2}-\epsilon}.$$

Theorem 3 *Let p be an odd prime. Then, for $p \equiv 3 \pmod 8$, we have*

$$Q(p, S_4) = \frac{1}{2}\left[\frac{p-1}{4}\right].$$

Also, for any $0 < \epsilon < \frac{1}{2}$, we have

$$Q(p, S_4) - \frac{p-1}{8} \gg_\epsilon p^{\frac{1}{2}-\epsilon}, \text{ if } p \equiv 1 \pmod 4,$$

and

$$Q(p, S_4) - \frac{1}{2}\left[\frac{p-1}{4}\right] \gg_\epsilon p^{\frac{1}{2}-\epsilon}; \text{ if } p \equiv 7 \pmod 8.$$

Corollary 1.3 *Let p be an odd prime. Then, for $p \equiv 3 \pmod 8$, we have*

$$N(p, S_4) = \frac{1}{2}\left[\frac{p-1}{4}\right].$$

Also, for any $0 < \epsilon < \frac{1}{2}$, we have

$$\frac{p-1}{8} - N(p, S_4) \gg_\epsilon p^{\frac{1}{2}-\epsilon}; \text{ if } p \equiv 1 \pmod 4,$$

and

$$\frac{1}{2}\left[\frac{p-1}{4}\right] - N(p, S_4) \gg_\epsilon p^{\frac{1}{2}-\epsilon}; \text{ if } p \equiv 7 \pmod 8.$$

Using Theorems 1 and 3, we conclude the following corollary.

Corollary 1.4 *Let p be an odd prime such that $p \equiv 3 \pmod 8$. Then for any ϵ with $0 < \epsilon < \frac{1}{2}$, we have*

$$Q(p, S_2 \backslash S_4) - \frac{1}{2}\left\lfloor\frac{p-1}{4}\right\rfloor \gg_\epsilon p^{\frac{1}{2}-\epsilon}.$$

2 Preliminaries

In this section, we shall state many useful results as follows.

Theorem 4 (Polya–Vinogradov) *Let p be any odd prime and χ be a non-principal Dirichlet character modulo p. Then, for any integers $0 \le M < N \le p-1$, we have*

$$\left| \sum_{m=M}^{N} \chi(m) \right| \le \sqrt{p} \log p.$$

Let us define the following counting functions as follows. Let

$$f(x) = \frac{1}{2}\left(1 + \left(\frac{x}{p}\right)\right) \text{ for all } x \in (\mathbb{Z}/p\mathbb{Z})^* \tag{2}$$

and

$$g(x) = \frac{1}{2}\left(1 - \left(\frac{x}{p}\right)\right) \text{ for all } x \in (\mathbb{Z}/p\mathbb{Z})^* \tag{3}$$

where $\left(\frac{\cdot}{p}\right)$ is the Legendre symbol. Then, we have

$$f(x) = \begin{cases} 1; \text{ if } x \text{ is a quadratic residue} \pmod{p}, \\ 0; \text{ otherwise.} \end{cases}$$

and

$$g(x) = \begin{cases} 1; \text{ if } x \text{ is a quadratic non-residue} \pmod{p}, \\ 0; \text{ otherwise.} \end{cases}$$

In the following lemma, we prove the "expected" result.

Lemma 1 *For an integer $k \ge 1$ and an odd prime p, let $S_k = kI$ where I is the interval $I = \{1, 2, \ldots, [(p-1)/k]\}$. Then*

$$Q(p, S_k) = \frac{1}{2}\left[\frac{p-1}{k}\right] + \frac{1}{2}\left(\frac{k}{p}\right) \sum_{m=1}^{(p-1)/k} \left(\frac{m}{p}\right) \tag{4}$$

and hence

$$Q(p, S_k) = \frac{1}{2}\left[\frac{p-1}{k}\right] + O(\sqrt{p}\log p).$$

The same expressions hold for $N(p, S_k)$ as well.

Proof We prove for $Q(p, S_k)$ and the proof of $N(p, S_k)$ follows analogously. Let ψ_k be the characteristic function for S_k which is defined as

$$\psi_k(m) = \begin{cases} 1; \text{ if } m \in S_k, \\ 0; \text{ if } m \notin S_k. \end{cases}$$

Now, by (2), we see that

$$Q(p, S_k) = \sum_{m \in S_k} f(m) = \sum_{m=1}^{p-1} \psi_k(m) f(m) = \frac{1}{2} \sum_{m=1}^{p-1} \psi_k(m) \left(1 + \left(\frac{m}{p}\right)\right)$$
$$= \frac{1}{2}\left[\frac{p-1}{k}\right] + \frac{1}{2}\left(\frac{k}{p}\right) \sum_{m=1}^{(p-1)/k} \left(\frac{m}{p}\right), \tag{5}$$

which proves (4). Then, by Theorem 4, we get

$$Q(p, S_k) = \frac{1}{2}\left[\frac{p-1}{k}\right] + O(\sqrt{p}\log p).$$

This finishes the proof. □

Let $q > 1$ be a positive integer and let ψ be a nontrivial quadratic character modulo q. Let $L(s, \psi) = \sum_{n=1}^{\infty} \frac{\psi(n)}{n^s}$ be the Dirichlet L-function associated to ψ. Since ψ is a nontrivial homomorphism, $L(s, \psi)$ admits the following Euler product expansion:

$$L(s, \psi) = \prod_{p \nmid q} \left(1 - \frac{\psi(p)}{p^s}\right)^{-1}$$

for all complex number s with $\Re(s) > 1$. This, in particular, shows that $L(s, \psi) > 0$ for all real number $s > 1$. By continuity, it follows that $L(1, \psi) \geq 0$. Dirichlet proved that $L(1, \psi) \neq 0$ in order to prove the infinitude of prime numbers in an arithmetic progression. Hence, it follows that $L(1, \psi) > 0$ for all nontrivial quadratic character ψ. Since $L(1, \psi) > 0$, it is natural to expect some nontrivial lower bound as a function of q. This is what was proved by Landau–Siegel in the following theorem. The proof can be found in [14].

Theorem 5 *Let $q > 1$ be a positive integer and ψ be a nontrivial quadratic character modulo q. Then for each $\epsilon > 0$, there exists a constant $C(\epsilon) > 0$ such that*

$$L(1, \psi) > \frac{C(\epsilon)}{q^{\epsilon}}.$$

The following lemma is crucial for our discussions. This lemma connects the sum of Legendre symbols and the Dirichlet L-function associated with Legendre symbol via the famous Dirichlet class number formula for the quadratic field. For an odd prime p, the Legendre symbol $\left(\frac{\cdot}{p}\right) = \chi_p(\cdot)$ is a quadratic Dirichlet character modulo p. We also define a character

$$\chi_4(n) = \begin{cases} (-1)^{(n-1)/2}; & \text{if } n \text{ is odd,} \\ 0; & \text{otherwise.} \end{cases}$$

Then one can define the Dirichlet character χ_{4p} as $\chi_{4p}(n) = \chi_4(n)\chi_p(n)$ for any odd prime p and similarly, we can define $\chi_{3p}(n) = \chi_3(n)\chi_p(n)$ for any odd prime $p > 3$. Clearly, χ_{4p} and χ_{3p} are nontrivial and real quadratic Dirichlet characters.

Lemma 2 *(See for instance, Page 151, Theorem 7.2 and 7.4 in [24]) Let $p > 3$ be an odd prime and for any real number $\ell \geq 1$, we define*

$$S(1,\ell) = \sum_{1 \leq m < \ell} \chi_p(m). \tag{6}$$

Then we have the following equalities.

(1) For a prime $p \equiv 3 \pmod 4$, we have

$$S(1, p/2) = \frac{\sqrt{p}}{\pi}\left(2 - \chi_p(2)\right) L(1, \chi_p),$$

where $L(1, \chi_p)$ is the Dirichlet L-function; Also, we have

$$S(1, p/3) = \frac{\sqrt{p}}{2\pi}(3 - \chi_p(3))L(1, \chi_p).$$

(2) For a prime $p \equiv 1 \pmod 4$, we have

$$S(1, p/3) = \frac{\sqrt{3p}}{2\pi} L(1, \chi_{3p});$$

Also, we have

$$S(1, p/4) = \frac{\sqrt{p}}{\pi} L(1, \chi_{4p}).$$

Now, we need the following lemma, which deals with the vanishing sums of Legendre symbols. This was proved in [2]. For more such relations one may refer to [8].

Lemma 3 *[2] Let p be an odd prime. Then the following equalities hold true.*

(1) If $p \equiv 1 \pmod 4$, then we have $\displaystyle\sum_{n=1}^{(p-1)/2} \left(\frac{n}{p}\right) = 0.$

(2) If $p \equiv 3 \pmod 8$, then we have $\displaystyle\sum_{n=1}^{\lfloor p/4 \rfloor} \left(\frac{n}{p}\right) = 0.$

(3) If $p \equiv 7 \pmod 8$, then we have $\displaystyle\sum_{\lceil p/4 \rceil}^{\lfloor p/2 \rfloor} \left(\frac{n}{p}\right) = 0.$

3 Proof of Theorem 1

Let p be a given odd prime. We want to estimate the quantity $Q(p, S_2)$. Therefore, by (5), we get

$$Q(p, S_2) = \frac{1}{2}\left[\frac{p-1}{2}\right] + \frac{1}{2}\left(\frac{2}{p}\right)\sum_{n=1}^{(p-1)/2}\left(\frac{n}{p}\right). \tag{7}$$

Now, we consider three cases as follows.

Case 1. $p \equiv 1 \pmod 4$

In this case, since $\sum_{n=1}^{(p-1)/2}\left(\frac{n}{p}\right) = 0$, by Lemma 3 (1), the Eq. (7) reduces to

$$Q(p, S_2) = \frac{p-1}{4},$$

which is as desired.

Case 2. $p \equiv 3 \pmod 8$

By Lemma 2 (1) and by (7), we get

$$Q(p, S_2) = \frac{1}{2}\left[\frac{p-1}{2}\right] + \frac{\sqrt{p}}{\pi}\left(2 - \chi_p(2)\right) L(1, \chi_p).$$

In this case, we know that $\left(\frac{2}{p}\right) = -1$. Therefore, we get

$$Q(p, S_2) = \frac{1}{2}\left[\frac{p-1}{2}\right] + 3\frac{\sqrt{p}}{\pi} L(1, \chi_p).$$

Let ϵ be any real number such that $0 < \epsilon < \frac{1}{2}$. Then by Theorem 5, we get

$$Q(p, S_2) - \frac{1}{2}\left[\frac{p-1}{2}\right] \gg_\epsilon p^{\frac{1}{2}-\epsilon},$$

as desired.

Case 3. $p \equiv 7 \pmod 8$.

Since $p \equiv 7 \pmod 8$, we know that $\left(\frac{2}{p}\right) = 1$. Therefore, by Lemma 2 (1) and by (7), we get

$$Q(p, S_2) = \frac{1}{2}\left[\frac{p-1}{2}\right] + \frac{\sqrt{p}}{\pi} L(1, \chi_p) = \frac{1}{2}\left[\frac{p-1}{2}\right] + \frac{\sqrt{p}L(1, \chi_p)}{\pi}.$$

Let ϵ be any real number such that $0 < \epsilon < \frac{1}{2}$. Then by Theorem 5 we get

$$Q(p, S_2) - \frac{1}{2}\left[\frac{p-1}{2}\right] \gg_\epsilon p^{\frac{1}{2}-\epsilon}$$

which proves the theorem. □

4 Proof of Theorem 2

Let p be a given odd prime. We want to estimate the quantity $Q(p, S_3)$. Therefore, by (5), we get,

$$Q(p, S_3) = \frac{1}{2}\left[\frac{p-1}{3}\right] + \left(\frac{3}{p}\right)\sum_{n=1}^{(p-1)/3}\left(\frac{n}{p}\right). \tag{8}$$

Now, we consider the following cases.

Case 1. $p \equiv 1 \pmod{12}$

Note that, in this case, we have $\left(\frac{3}{p}\right) = 1$. By (8) and by Lemma 2 (2), we get

$$\begin{aligned} Q(p, S_3) - \frac{1}{2}\left(\frac{p-1}{3}\right) &= \frac{1}{2}\frac{\sqrt{3p}}{2\pi}L(1, \chi_3\chi_p) \\ &\geq \frac{\sqrt{3p}}{4\pi}\frac{C(\epsilon)}{(3p)^\epsilon} \\ &\gg_\epsilon p^{\frac{1}{2}-\epsilon}, \end{aligned}$$

for any given $0 < \epsilon < \frac{1}{2}$ in Theorem 5.

Case 2. $p \equiv 11 \pmod{12}$

In this case, we have, $\left(\frac{3}{p}\right) = 1$. Then again by (8) and by Lemma 2 (1), we get

$$Q(p, S_3) = \frac{1}{2}\left[\frac{p-1}{3}\right] + \frac{1}{2}\frac{\sqrt{3p}}{2\pi}(3 - \chi_p(3))L(1, \chi_p).$$

Hence

$$Q(p, S_3) - \frac{1}{2}\left[\frac{p-1}{3}\right] \gg_\epsilon p^{\frac{1}{2}-\epsilon},$$

for any $0 < \epsilon < \frac{1}{2}$ in Theorem 5. □

5 Proof of Theorem 3

At first, using the Eq. (5), we note that

$$Q(p, S_4) = \frac{1}{2}\left[\frac{p-1}{4}\right] + \frac{1}{2}\left(\frac{4}{p}\right)\sum_{m=1}^{(p-1)/4}\left(\frac{m}{p}\right) = \frac{1}{2}\left[\frac{p-1}{4}\right] + \frac{1}{2}\sum_{m=1}^{(p-1)/4}\left(\frac{m}{p}\right). \tag{9}$$

Case 1. $p \equiv 1 \pmod 4$

Now, we apply Lemma 2 (2) in (9) and we get

$$Q(p, S_4) = \frac{1}{2}\left(\frac{p-1}{4}\right) + \frac{1}{2}\frac{\sqrt{p}}{\pi}L(1, \chi_4\chi_p).$$

Hence

$$Q(p, S_4) - \frac{p-1}{8} \gg_\epsilon p^{\frac{1}{2}-\epsilon},$$

for any $0 < \epsilon < \frac{1}{2}$ in Theorem 5.

Case 2. $p \equiv 3 \pmod 8$

In this case, we apply Lemma 3 (2) which says that $\sum_{n=1}^{[(p-1)/4]}\left(\frac{m}{p}\right) = 0$. Hence, by (9), we get

$$Q(p, S_4) = \frac{1}{2}\left[\frac{p-1}{4}\right].$$

Case 3. $p \equiv 7 \pmod 8$

First note that by Lemma 3 (3), we have

$$\sum_{\frac{p-1}{4} < m < \frac{p-1}{2}}\left(\frac{m}{p}\right) = 0.$$

Therefore, the Eq. (9) can be rewritten as

$$\begin{aligned}Q(p, S_4) &= \frac{1}{2}\left[\frac{p-1}{4}\right] + \frac{1}{2}\sum_{1\le m\le (p-1)/4}\left(\frac{m}{p}\right) + \frac{1}{2}\sum_{(p-1)/4\le m\le (p-1)/2}\left(\frac{m}{p}\right)\\ &= \frac{1}{2}\left[\frac{p-1}{4}\right] + \frac{1}{2}\sum_{m=1}^{\frac{p-1}{2}}\left(\frac{m}{p}\right).\end{aligned}$$

Now, by Lemma 2 (1), we get

$$Q(p, S_4) = \frac{1}{2}\left[\frac{p-1}{4}\right] + \frac{1}{2}\frac{\sqrt{p}}{\pi}L(1, \chi_p).$$

Hence

$$Q(p, S_4) - \frac{1}{2}\left[\frac{p-1}{4}\right] \gg_\epsilon p^{\frac{1}{2}-\epsilon},$$

for any $0 < \epsilon < \frac{1}{2}$ in Theorem 5. This proves the result. □

Acknowledgements We thank Professor V. Kumar Murty for going through the manuscript very carefully and for a suggestion to clear our doubts.

References

1. A. Brauer, Űber Sequenzen von Potenzresten. Sitzungsberichte der Preubischen Akademie der Wissenschaften 9–16 (1928)
2. B.C. Berndt, S. Chowla, Zero sums of the legendre symbol. Nordisk Mat. Tidskr. **22**, 5–8 (1974)
3. L. Carlitz, Sets of primitive roots. Compos. Math. **13**, 65–70 (1956)
4. A. Gica, Quadratic residues of certain types. Rocky Mt. J. Math. **36**, 1867–1871 (2006)
5. S. Gun, B. Ramakrishnan, B. Sahu, R. Thangadurai, Distribution of quadratic non-residues which are not primitive roots. Math. Bohem. **130**(4), 387–396 (2005)
6. S. Gun, F. Luca, P. Rath, B. Sahu, R. Thangadurai, Distribution of residues modulo p. Acta Arith. **129**(4), 325–333 (2007)
7. M. Hausman, Primitive roots satisfying a coprime condition. Amer. Math. Monthly **83**, 720–723 (1976)
8. W. Johnson, K.J. Mitchell, Symmetries for sums of the legendre symbol. Pac. J. Math. **59**(1), 117–124 (1977)
9. B. Karaivanov, T.S. Vassilev, On certain sums involving the Legendre symbol. Integers **16**, A14 (2016)
10. W. Kohnen, An elementary proof of the theory of quadratic residues. Bull. Korean Math. Soc. **45** (2), 273–275 (2008)
11. A. Laradji, M. Mignotte, N. Tzanakis, Elementary trigonometric sums related to quadratic residues. Elem. Math. **67**, 51–60 (2012)
12. F. Luca, I.E. Shparlinski, R. Thangadurai, Quadratic non-residue verses primitive roots modulo p. J. Ramanujan Math. Soc. **23**(1), 97–104 (2008)
13. F. Luca, R. Thangadurai, Distribution of Residues Modulo p - II, to appear in the Ramanujan Mathematical Society Lecture Notes Series (2011)
14. H. Montgomery, R. Vaughan, *Multiplicative Number Theory I Classical Theory* (Cambridge University Press, Cambridge, 2007)
15. L. Moser, On the equation $\phi(n) = \pi(n)$. Pi Mu Epsilon J. 101–110 (1951)
16. P. Pollack, The least prime quadratic non-residue in a prescribed residue class mod 4. J. Number Theory **187**, 403–414 (2018)
17. M. Szalay, On the distribution of the primitive roots mod p (in Hungarian). Mat. Lapok **21**, 357–362 (1970)
18. M. Szalay, On the distribution of the primitive roots of a prime. J. Number Theory **7**, 183–188 (1975)
19. J. Tanti, R. Thangadurai, Distribution of residues and primitive roots. Proc. Indian Acad. Sci. **123**(2), 203–211 (2013)

20. E. Vegh, Primitive roots modulo a prime as consecutive terms of an arithmetic progression. J. Reine Angew. Math. **235**, 185–188 (1969)
21. E. Vegh, Arithmetic progressions of primitive roots of a prime. II, ibid. **244**, 108–111 (1970)
22. E. Vegh, A note on the distribution of the primitive roots of a prime. J. Number Theory **3**, 13–18 (1971)
23. E. Vegh, Arithmetic progressions of primitive roots of a prime. III. J. Reine Angew. Math. **256**, 130–137 (1972)
24. S. Wright, Quadratic residues and non-residues: selected topics. *Lecture notes in Mathematics*, vol. 2171. (Springer, 2016)
25. A. Weil, On the Riemann hypothesis. Proc. Nat. Acad. Sci. USA **27**, 345–347 (1941)

On Class Number Divisibility of Number Fields and Points on Elliptic Curves

Debopam Chakraborty

1 Introduction

The class group of a number field K measures how far its ring of integers is from having unique factorization into irreducible elements. It is the quotient of the group of all fractional ideals of K by the subgroups of principal fractional ideals. It is well known from class field theory that the ideal class group is also the Galois group of the maximal unramified abelian extension of K.

The class number problem originated even before the concept of ideal was discovered. It came from the work of Legendre and Euler in quadratic forms. Later in 1801, Gauss proposed three conjectures regarding class number of quadratic number fields in his book "Disquisitiones Arithmeticae". Two of them were about number fields with negative discriminant and they have been completely answered through contribution of several mathematicians, most notably Hecke, Deuring, and Heilbronn in 1930. For real quadratic fields, Gauss conjectured that there will be infinitely many real quadratic fields with class number as one. This is still an open problem and not much progress has been made regarding this conjecture till now.

2 Class Number Related Questions

Difficulty in solving the Gauss' Conjecture for Real Quadratic Fields, leads mathematicians to turn their focus on several related class number questions for number fields such as investigating class number divisibility for number fields of smaller degree, class number of a number field and its relation with fundamental unit, points

D. Chakraborty (✉)
Department of Mathematics, BITS-Pilani, Hyderabad Campus 500078, Telangana, India
e-mail: debopam@hyderabad.bits-pilani.ac.in

K. Chakraborty et al. (eds.), *Class Groups of Number Fields and Related Topics*,
https://doi.org/10.1007/978-981-15-1514-9_10

on an elliptic curve, and the behavior of the class number of number field generated from that point.

Soleng [5] gave a construction of families of quadratic number fields from an elliptic curve having ideal class group isomorphic to the torsion group of the curve. A. Sato constructed quadratic number fields with class number divisible by 5 from elliptic curves in [4]. Lemmermeyer [3] showed a method for constructing unramified quadratic extension of cubic fields using points on suitable elliptic curves.

The main focus of the talk is to give a glimpse of aforementioned work of Soleng and also to present one original work of the speaker [2] (a joint work with Prof. Anupam Saikia) which was inspired by the aforementioned work of F. Lemmermeyer.

3 Homomorphisms from the Group of Rational Points on Elliptic Curves to Class Group of Number Fields

In his 1994 paper [5], R. Soleng was looking to find the correspondence between rational point on an elliptic curve and nontrivial ideal classes of number fields. To find the answers of the previously mentioned questions, Soleng has started with defining a particular subset of $E(\mathbb{Q})$, named as set of "primitive points" defined as follows:

Definition 3.1 Let $P = (x, y)$ be a rational point on the elliptic curve $E : Y^2 = X^3 + a_2X^2 + a_4X + a_6$ defined over $\mathbb{Z}$. The point P is said to be "*primitive*" if $gcd(x, 2y, x^2 + a_2x + a_4) = 1$.

The point at infinity is considered as primitive by convention.

Then using the following two results, he explicitly gave a homomorphism from a subgroup of the Mordell–Weil group of $E(\mathbb{Q})$ to the ideal class group of $\mathbb{Q}(\sqrt{a_6})$.

Proposition 3.2 *Let E be an elliptic curve defined over the rational integers by the equation* $Y^2 = x^3 + a_2X^4 + a_4X + a_6$, *then the subset of* $E(\mathbb{Q})$ *consisting of primitive points is a group.*

Theorem 3.3 (Soleng) *Let E be an elliptic curve defined over* $\mathbb{Z}$ *by the equation*

$$Y^2 = x^3 + a_2X^2 + a_4X + a_6$$

and let $E(\mathbb{Q})_{prim}$ *be the subset of the Mordell–Weil group consisting of primitive points, then the map*

$$P = (\frac{A}{C^2}, \frac{B}{C^3}) \to (A, -kB + \sqrt{a_6}),$$

where k is an integer satisfying $kC^3 + lA = 1$ *for some* $l \in \mathbb{Z}$, *is a homomorphism from the group of primitive rational points on the curve to the ideal class group of the order* $\mathbb{Z} + \mathbb{Z}\sqrt{a_6}$ *in* $\mathbb{Q}(\sqrt{a_6})$.

The previously mentioned theorem has many interesting consequences that have been noted by the author himself in the same paper. It gives an affirmative answer to a conjecture of *Yves Hellegourach*.

4 A Construction for Biquadratic Fields of Even Class Number

As mentioned earlier, Lemmermeyer [3] showed a method for constructing unramified quadratic extension of cubic fields using points on suitable elliptic curves.

The genus field of $\mathbb{Q}(\sqrt{a}, \sqrt{b})$ has been discussed in detail by Yi and Zhe [6], Bae and Yue [1], and Yue [7] when at least one of a and b is a prime of the form 1 *mod* 4. Here we present quadratic unramified extensions of infinitely many biquadratic fields $\mathbb{Q}(\sqrt{r}, \sqrt{m})$ where, if m is suitably chosen, both r and m will be composite or none of r and m will be a prime congruent to 1 mod 4. The construction can be applied for infinitely many biquadratic fields $\mathbb{Q}(\sqrt{r_i}, \sqrt{3})$ where r_i's are square-free composite numbers. The main theorem is as follows.

Theorem 4.1 *Let $m \neq 0, 1$ be a square-free integer which is divisible by 3 if it is positive. Let $P_0 = \left(\frac{r_0}{t_0^2}, \frac{s_0}{t_0^3}\right)$ be any non-torsion point of the elliptic curve $y^2 = x^3 + m$ such that r_0 is odd and non-square. Let $\left(\frac{r_i}{t_i^2}, \frac{s_i}{t_i^3}\right) = 2^i P_0$ for each natural number i. Then the biquadratic field $K_i = \mathbb{Q}(\sqrt{r_i}, \sqrt{m})$ has an everywhere unramified quadratic extension $K_i(\sqrt{\beta_i})$, where β_i is either $\pm(s_i + t_i^3\sqrt{m})$ or $3(s_i + t_i^3\sqrt{m})$.*

The proof of the main theorem depends on the result of the following lemmas:

Lemma 4.2 *Consider the duplication formula for the point $P = (\frac{r}{t^2}, \frac{s}{t^3})$ on $y^2 = x^3 + m$:*

$$\left(\frac{r(2P)}{t(2P)^2}, \frac{s(2P)}{t(2P)^3}\right) = 2P = \left(\frac{r(9r^3 - 8s^2)}{(2st)^2}, \frac{27r^6 - 36r^3s^2 + 8s^4}{(2st)^3}\right) \tag{1}$$

Suppose m is square-free and r is odd. If $3 \nmid s$, the fractions on the right-hand side are already in their reduced form. When $s = 3s'$ the fractions on the right-hand side above reduces to

$$\left(\frac{r(2P)}{t(2P)^2}, \frac{s(2P)}{t(2P)^3}\right) = 2P = \left(\frac{r(r^3 - 8s'^2)}{(2s't)^2}, \frac{r^6 - 12r^3s'^2 + 24s'^4}{(2s't)^3}\right). \tag{2}$$

Lemma 4.3 *Let $P = \left(\frac{r}{t^2}, \frac{s}{t^3}\right)$ be a non-torsion point of the elliptic curve E_m satisfying certain conditions with t even. Then α and its conjugate $\bar{\alpha}$ over $\mathbb{Q}(\sqrt{r})$ generate coprime ideals in the ring $\mathcal{O}_K$ of integers in K. Moreover, there exists an ideal $\mathfrak{a}$ in $\mathcal{O}_K$ such that $\langle\alpha\rangle := \alpha\mathcal{O}_K = \mathfrak{a}^2$.*

Lemma 4.4 *The extension $K(\sqrt{\beta})$ over $K = \mathbb{Q}(\sqrt{r}, \sqrt{m})$ is quadratic and unramified at all finite primes.*

Lemma 4.5 *The infinite primes do not ramify in $K(\sqrt{\beta})/K$.*

It can also be shown that one can start with a non-torsion point P_0 and repeat the procedure of the previous section for each multiple $P_i = 2^i(P_0) = \left(\frac{r_i}{t_i^2}, \frac{s_i}{t_i^3}\right)$.

References

1. S. Bae, Q. Yue, Hilbert genus fields of real biquadratic fields. Ramanujan J. **24**, 161–181 (2011)
2. D. Chakraborty, A. Saikia, An explicit construction for unramified quadratic extensions of biquadratic fields. Acta Arithmetica. **178**, 153–161 (2017)
3. F. Lemmermeyer, Why the class number of $\mathbb{Q}(\sqrt[3]{11})$ is even? Math. Bohem. **138**(2), 149–163 (2013)
4. A. Sato, On the class numbers of certain number fields obtained from points on elliptic curves III. Osaka. J. Math. **48**, 809–826 (2011)
5. R. Soleng, Homomorphisms from the group of rational points on elliptic curves to class groups of quadratic number fields. J. Number Theory **46**, 214–229 (1994)
6. O. Yi, Z. Zhe, Hilbert genus fields of biquadratic fields. Sci. China Math. **57**, 2111–2122 (2014)
7. Q. Yue, Genus fields of real biqudratic fields. Ramanujan J. **21**, 17–25 (2010)

Small Fields with Large Class Groups

Florian Luca and Preda Mihăilescu

1 Introduction

It is a folklore expectation, that for a given abelian group G, one can construct quadratic fields $\mathbb{K}/\mathbb{Q}$ which class group $\mathcal{C}(\mathbb{K}) = G$. Many papers have been published in the literature showing how one can construct fields $\mathbb{K}$ of a given degree d over $\mathbb{Q}$ whose class groups contain a copy of H, where H is some given abelian group, usually of small rank (that is, a group which can be realised as a subgroup of $(\mathbb{Z}/n\mathbb{Z})^r$ for some small r). We mention papers by Mestre [5], Bilu and Luca [1], Levin [4] and Bilu and Gillibert [2]. For example, Mestre [5] showed that there are infinitely many quadratic $\mathbb{K}$ such that $\mathcal{C}(\mathbb{K})$ contains a copy of $(\mathbb{Z}/5\mathbb{Z})^3$, while Bilu and Gillibert [2] have extended Mestre's result. In private discussions of the second author at the Harish-Chandra Research Institute, Francois Biasse mentioned that having a simple construction of small degree fields with large class groups might be useful in cryptography. Another interesting connection relates class groups to so-called *algebraic lattices*, with numerous applications in coding and cryptography. However, recent research showed that algebraic lattices do not, in general, behave like random lattices with respect to the shortest vector problem (SVP), which is easier to solve in the case of classes from the minus part of abelian extensions, where Stickleberger

F. Luca
School of Mathematics, University of the Witwatersrand,
Private Bag X3, Wits 2050, South Africa
e-mail: florian.luca@wits.ac.za

Max Planck Institute for Mathematics, Vivatsgasse 7, 53111 Bonn, Germany

Faculty of Sciences, Department of Mathematics, University of Ostrava,
30 Dubna 22, 701 03 Ostrava 1, Czech Republic

P. Mihăilescu (✉)
Mathematisches Institut der Universität Göttingen, Göttingen, Germany
e-mail: preda@uni-math.gwdg.de

K. Chakraborty et al. (eds.), *Class Groups of Number Fields and Related Topics*,
https://doi.org/10.1007/978-981-15-1514-9_11

annihilators can be used [3]. Our constructions work in CM fields, so the Stickelberger attack can be avoided.

With this motivation, in this paper, we offer the following result:

Theorem 1 *Let $\mathbb{K}$ be a CM extension of $\mathbb{Q}$. Let p be an odd prime and n a positive integer. Then there is an abelian extension $\mathbb{L}/\mathbb{K}$ of degree p such that $\mathcal{C}(\mathbb{L})$ contains a copy of $(\mathbb{Z}/p\mathbb{Z})^n$.*

A related result is given by Washington in [6] Proposition 3.8, and the proof uses different arguments from ours.

2 Proof of the Theorem

Throughout this paper, all the fields denoted $\mathbb{K}$ and $\mathbb{L}$ are Galois and CM, that is quadratic imaginary extensions of a totally real extension of $\mathbb{Q}$. Let $\mathbb{K}$ be an arbitrary Galois CM field and $Q = \{q_i : i = 1, 2, \ldots, n\}$ be a set of n primes such that q_i is totally split in $\mathbb{K}$ and $q_i \equiv 1 \pmod{p}$. Put $N = \prod_{1 \le i \le n} q_i$. We will take $\mathbb{L} \supset \mathbb{K}$ to be a suitable degree p extension of $\mathbb{K}$, which we will describe shortly.

First, some notation. The global complex conjugation of the field $\mathbb{C}$ embeds canonically in $\mathrm{Gal}(\mathbb{L}/\mathbb{Q})$, and we denote this embedding by $\jmath \in \mathrm{Gal}(\mathbb{L}/\mathbb{Q})$. Complex conjugation acts naturally on various groups attached to $\mathbb{L}$. For example, it induces a canonical decomposition of the group of principal ideals $P(\mathbb{L})$ of $\mathbb{L}$ and therefore of the class group $\mathcal{C}(\mathbb{L})$ and in particular of the p-Sylow group denoted $A(\mathbb{L}) = (\mathcal{C}(\mathbb{L}))_p$, in minus and class parts. In the instance of the principal ideal group, these are given by

$$P(\mathbb{L})^+ = \{(\alpha)^{\jmath+1} : (\alpha) \in P(\mathbb{L})\}, \ P(\mathbb{L})^- = \{(\alpha)^{1-\jmath} : (\alpha) \in P(\mathbb{L})\}.$$

That is, $P(\mathbb{L})^+$ and $P(\mathbb{L})^-\}$ are principal ideals of the form $(\alpha\bar{\alpha})$ and $\alpha/\bar{\alpha}$, respectively. Clearly, $P(\mathbb{L})^2 = P(\mathbb{L})^- \cdot P(\mathbb{L})^+$, and similarly for all other groups. Since p is odd, we have that $A(\mathbb{L}) = A(\mathbb{L})^{p^k+1}$ for some large enough k, therefore we deduce that in fact

$$A(\mathbb{L}) = A(\mathbb{L})^{p^k+1} = A^+(\mathbb{L}) \cdot A^-(\mathbb{L}).$$

For $m \in \mathbb{N}$, we denote by ζ_m a primitive m-th roots of unity and by $\mathbf{C}_m = \mathbb{Q}[\zeta_m]$ the m-th cyclotomic extension. Recall that $\mathbf{C}_m, \mathbf{C}_{m'}$ are linearly disjoint over $\mathbb{Q}$ whenever $\gcd(m, m') = 1$.

Lemma 1 *Let $\mathbf{C}_N = \mathbb{Q}[\zeta_N]$. Then there is a subfield $\mathbb{F} \subset \mathbf{C}_N$ of degree p over $\mathbb{Q}$ such that all the primes q_i are ramified in $\mathbb{F}/\mathbb{Q}$ for all $i = 1, \ldots, n$.*

Proof Induction on n. For $n = 1$, the index N is prime and $\mathbf{C}_N$ is totally ramified at q and it contains a unique extension of degree p over $\mathbb{Q}$; the claim follows by letting $\mathbb{F}$ be this extension. Assume next that $n > 1$. Let $\mathbb{M} \subset \mathbf{C}_N$ be the maximal p-elementary subextension of $\mathbb{Q}$ in $\mathbf{C}_N$. Thus, $\mathrm{Gal}(\mathbb{M}/\mathbb{Q})^p = 1$ and $\mathbb{M}$ is maximal with this property. Then $\mathbb{M}$ is the compositum of n cyclic fields of degree p, obtained for each of the primes q_i as in the case $n = 1$. In particular, $G_N := \mathrm{Gal}(\mathbb{M}/\mathbb{Q}) \cong (\mathbb{Z}/p\mathbb{Z})^n$.

When $n = 2$, the inertia group $I(q_i) \subset G_N$ of each of the two primes q_1 and q_2 fixes one of the $\frac{p^2-1}{p-1} = p + 1$ subfields of degree p in $\mathbf{C}_N$. Consequently, there are $(p+1) - 2 = p - 1 > 0$ subfields of degree p in $\mathbb{M}$, in which both primes ramify. For $n > 2$, the inertia groups $I(q_i)$ fix subfields with Galois groups of p-rank equal to $n - 1$. A simple combinatorial argument implies that there are $\frac{p^n-1}{p-1}$ subfields of $\mathbb{M}$, which are fixed by a cyclic group of order p. Out of these, only n of them are fixed by some inertia group $I(q_i)$ for some $i \in \{1, \ldots, n\}$. Thus, there are $\frac{p^n-1}{p-1} - n > 0$ fields in which all the q_i's are totally ramified, and we choose $\mathbb{F}$ in any of these subfields. □

We now let $\mathbb{K}$ be an arbitrary Galois CM field, Q be a list of n primes as above, and $\mathbb{F} \subset \mathbf{C}_N$ be cyclic extension of degree p over $\mathbb{Q}$ in which all these primes are ramified, which exists by Lemma 1. We let $\mathbb{L} = \mathbb{K}\mathbb{F}$. This is an abelian extension of degree p of $\mathbb{K}$, for otherwise $\mathbb{F}$ will be a subfield of $\mathbb{K}$; this is false, since the q_i's ramify in $\mathbb{F}$ but not in $\mathbb{K}$ (being split in $\mathbb{K}$, by hypothesis). Let $\Phi = \mathrm{Gal}(\mathbb{F}/\mathbb{Q}) = \mathrm{Gal}(\mathbb{L}/\mathbb{K})$ be generated by σ, so $\Phi = \langle\sigma\rangle$. We write $\mathcal{N} = \sum_{i=0}^{p-1} \sigma^i$ for the algebraic norm attached to this group and $s = \sigma - 1$ for a generator of its augmentation. The next fact follows from the Hasse Norm Principle.

Lemma 2 *There is an injective homomorphism*

$$\psi : P(\mathbb{K})^-/\mathcal{N}(P(\mathbb{L})^-) \hookrightarrow A^-(\mathbb{L}). \tag{1}$$

Let $\mathbf{R} = (\mathbb{Z}/p\mathbb{Z})^{[\mathbb{K}:Q]/2}$ *and*

$$\delta = \begin{cases} 1 & \text{if } \zeta_p \in \mathbb{K}; \\ 0 & \text{otherwise.} \end{cases}$$

Then there is a map of $\mathbb{F}_p$*-modules,*

$$\kappa : \mathbf{R}^n/(\mathbb{Z}/p\mathbb{Z})^\delta \hookrightarrow P(\mathbb{K})^-/\mathcal{N}(P(\mathbb{L})^-),$$

such that the map

$$\psi \circ \kappa : \mathbf{R}^n/(\mathbb{Z}/(p \cdot \mathbb{Z}))^\delta \hookrightarrow A^-(\mathbb{L}) \tag{2}$$

is an injection. In particular, the p-rank of $A^-(\mathbb{K})$ *is at least as large as* $n \cdot [\mathbb{K}^+ : \mathbb{Q}] - \delta$.

Proof The group $\mathbb{K}^\times/\mathcal{N}(\mathbb{L}^\times)$ has exponent p, since for $x \in \mathbb{K}^\times \subset \mathbb{L}^\times$ we have $x^p = \mathcal{N}(x) \in \mathcal{N}(\mathbb{L}^\times)$. Thus, $\mathbf{N} := P(\mathbb{K})^-/\mathcal{N}(P(\mathbb{L})^-)$ has exponent p. We define an injection

$$\psi : \mathbf{N} \to \mathrm{Ker}(\mathcal{N} : A^-(\mathbb{L}) \to A^-(\mathbb{K}))$$

as follows. Let $a \in A^-(\mathbb{L})$ have norm $\mathcal{N}(a) = 1$ and let $\mathfrak{A} \in a$ be some ideal. Then $\mathcal{N}(\mathfrak{A}) = (\alpha) \in P(\mathbb{K})$ and $\mathcal{N}(\mathfrak{A}^{1-\jmath}) = (\alpha/\bar{\alpha})$ is a principal ideal whose generator is uniquely defined up to roots of unity. Indeed, suppose that $(\alpha) = (\beta)$ and thus $(\alpha^{1-\jmath}) = (\beta^{1-\jmath})$ as principal ideals. Then there is a unit $\varepsilon \in \mathbb{K}$ such that $\alpha^{1-\jmath} = \varepsilon\beta^{1-\jmath}$ as algebraic numbers. By multiplying with the complex conjugate, we obtain $\varepsilon^{1+\jmath} = 1$. The Kronecker Unit Theorem implies that ε must be a root of unity. If $\zeta_p \notin \mathbb{K}$, then $\alpha^{1-\jmath}$ is the unique generator of its principal ideal, otherwise all the generators are of the form $\zeta_p^k\alpha^{1-\jmath}$ for $k \in \{0, 1, \ldots, p-1\}$. In this case, we have a surjective morphism $\iota : (\mathbb{K}^\times)^- \to P^-(\mathbb{K})$ with kernel $\langle\zeta_p\rangle$. Note also, that $\mathfrak{A}^\jmath \in a^{-1}$ since $a \in A^-(\mathbb{L})$, therefore $\mathfrak{A}^{1-\jmath} \in a^2$. If $(\alpha^{1-\jmath}) \in \mathcal{N}(P^-(\mathbb{L}))$, then there is an ideal $B \in P^-(\mathbb{L})$ such that $\mathcal{N}(B) = (\alpha^{1-\jmath})$. Hence, $\mathcal{N}(\mathfrak{A}^{1-\jmath}/B) = 1$, and since $\mathrm{Ker}(\mathcal{N} : P(\mathbb{L}) \to P(\mathbb{K})) = P^s(\mathbb{L})$, it follows in particular that $a \in (A^-(\mathbb{L}))^s$. Suppose that $(\alpha^{1-\jmath}) \notin \mathcal{N}(P^-(\mathbb{L}))$. Then, for any $\mathfrak{B} \in a$, there is some $B \in P(\mathbb{L})$ with $B \cdot \mathfrak{B} = \mathfrak{A}$, and thus

$$\mathcal{N}(\mathfrak{B}^{1-\jmath}) \in (\alpha^{1-\jmath}) \cdot \mathcal{N}(P^-(\mathbb{L})).$$

To the class $\mathbf{a} = (\alpha^{1-\jmath}) \cdot \mathcal{N}(P^-(\mathbb{L})) \in \mathbf{N}$, we have thus a well-defined association $\psi(\mathbf{a}) = a^2 \in A^-(\mathbb{L})$, since the previous argument shows that it does not depend on the choice of $(\alpha^{1-\jmath}) \cdot \mathcal{N}(P^-(\mathbb{L}))$. The same argument also shows injectivity, since we have seen that all ideals $\mathfrak{B} \in a$ have image $\mathcal{N}(\mathfrak{B}^{1-\jmath}) \in \mathbf{a}$. The arguments given above actually suffice for showing that ψ is an isomorphism

$$\psi : \mathbf{N} \to \hat{H}^1(\Phi, A^-(\mathbb{L})),$$

where $\hat{H}^1$ is the Tate cohomology group. This fact is irrelevant to our context.

It remains to estimate the size of $\mathbf{N}$. The Hasse Norm Principle implies that if $A = (\alpha^{1-\jmath}) \in \mathcal{N}(P^-(\mathbb{L}))$, then $\alpha^{1-\jmath}$ must be a norm at all ramified primes.

Let $q \in Q$. By construction q ramifies in $\mathbb{L}/\mathbb{K}$. We assumed it to be totally split in $\mathbb{K}/\mathbb{Q}$, so let $\mathfrak{q} \subset \mathbb{K}$ be a prime above q. Then all the other primes above q are conjugate under $G_{\mathbb{K}} = \mathrm{Gal}(\mathbb{K}/\mathbb{Q})$, and the completion $\mathbb{K}_q$ is isomorphic to $\mathbb{Q}_q$, the field of q-adic rationals. If $\mathfrak{Q} \subset \mathbb{L}$ is the ramified prime above q, then $\mathbb{L}_{\mathfrak{Q}}/\mathbb{K}_q$ is the subfield of degree p in the ramified qth cyclotomic extension of $\mathbb{Q}_q$, which is isomorphic to $\mathbb{F}_{\mathfrak{Q}}$ by restriction of primes. Local class field theory teaches us that

$$\mathbb{Q}_q/\mathcal{N}(\mathbb{F}_{\mathfrak{Q}}) \cong \mathrm{Gal}(\mathbb{F}_{\mathfrak{Q}}/\mathbb{Q}_q) \cong \mathrm{Gal}(\mathbb{L}_{\mathfrak{Q}}/\mathbb{K}_q),$$

under the local Artin symbol. Since the extension is tamely ramified, the norm defect is a subgroup of the roots of unity of order $q-1$ denoted $W \subset \mathbb{Z}_q^\times$. More precisely, there is $g \in \mathbb{Z}$ with $g^{(q-1)/p^{l-1}} \not\equiv 1 \pmod{q}$, where $l = v_p(q-1)$, such that for any $x \in \mathbb{Z}_q$, we have $x \notin \mathcal{N}(\mathbb{F}_{\mathfrak{Q}})$ if and only if $x \in g^c \cdot \mathcal{N}(\mathbb{F}_{\mathfrak{Q}})$ for some integer c coprime to p.

Let $A = (\alpha^{1-\jmath}) \in \mathbf{a} \in \mathbf{N}$ be a principal ideal and suppose that $\alpha^{1-\jmath} \equiv c \pmod{\mathfrak{q}}$. Then $1/\alpha^{1-\jmath} \equiv c \pmod{\bar{\mathfrak{q}}}$. Let $\mathfrak{q}^+ = \mathfrak{q} \cdot \bar{\mathfrak{q}}$ and $\theta_q \in \mathcal{O}(\mathbb{K})/\mathfrak{q}^+$ be the unique residue class verifying $\theta_q \equiv c \pmod{\mathfrak{q}}$ and $\theta_q \equiv c \pmod{\bar{\mathfrak{q}}}$. This defines an injective morphism

$$\theta_q : \mathbb{Z}/p\mathbb{Z} \hookrightarrow \mathcal{O}(\mathbb{K})/\mathfrak{q}^+.$$

For each $\tau \in \mathrm{Gal}(\mathbb{K}^+/\mathbb{Q})$, one defines in a similar way a morphism $\theta_{\tau(q)}$: $\mathbb{Z}/p\mathbb{Z} \hookrightarrow \mathcal{O}(\mathbb{K})/\tau(\mathfrak{q}^+)$. By the Chinese Remainder Theorem, we have

$$\mathcal{O}(\mathbb{K})/q\mathcal{O}(\mathbb{K}) \cong \prod_{\tau \in \mathrm{Gal}(\mathbb{K}^+/\mathbb{Q})} \mathcal{O}(\mathbb{K})/\tau(\mathfrak{q}^+).$$

We define $\theta : \mathbf{R} \hookrightarrow \mathcal{O}(\mathbb{K})/q\mathcal{O}(\mathbb{K})$ as the product of the maps θ_q mapping into $\mathcal{O}(\mathbb{K})/q\mathcal{O}(\mathbb{K})$ via the Chinese Remainder Theorem decomposition. Thus, for $\mathbf{c} = (c_\tau)_{\tau \in \mathrm{Gal}(\mathbb{K}^+/\mathbb{Q})} \in \mathbf{R}$, we let

$$\theta(\mathbf{c}) \equiv \theta_{\tau(q)}(c_\tau) \mod \tau(\mathfrak{q}^+).$$

By construction, if $\alpha^{1-j} \pmod{}\mathfrak{q}^+ \in \theta(\mathbf{R} \setminus \{0\}$, then $A \notin \mathcal{N}(\mathbb{L}^\times)$. The construction can be performed for each prime $q \in Q$ and we obtain a map

$$\theta : \mathbf{R} \hookrightarrow \mathcal{O}(\mathbb{K})/(N\mathcal{O}(\mathbb{K})),$$

which can be extended to a map $\bar{\theta} : \mathbf{R} \to \mathbf{N}$ in the natural way. Since the map ι : $(\mathbb{K}^\times)^- \to P^-(\mathbb{K})$ has the kernel $\langle \zeta_p \rangle$, we obtain $\kappa = \iota \circ \theta$, which is by construction injective, since we factor $\mathbf{R}$ out by $\mathbb{Z}/(p \cdot \mathbb{Z})^\delta$, which is the factor corresponding to the kernel. Finally, by combining κ with ψ, we see that (2) holds as claimed, which completes the proof. □

Theorem 1 is a direct consequence of Lemma 2. Note that if $\mathbb{K}$ is an imaginary quadratic extension and $p > 3$ or $p = 3$ and $\zeta_3 \notin \mathbb{K}$, we still have $(\mathbb{Z}/p\mathbb{Z})^n \hookrightarrow A^-(\mathbb{L}) \subset A(\mathbb{L})$. Also, if $\zeta_p \in \mathbb{K}$ and thus $\delta = 1$, we also have $[\mathbb{K}^+ : \mathbb{Q}] \geq \frac{p-1}{2} > 1$ for $p > 3$, and thus we construct a class group of p-rank at least $2n - 1$ by using the n ramified primes. If $p = 3$ and $\mathbb{K}^+ = \mathbb{Q}$, we need to assume $\mathbb{K} \neq \mathbb{Q}[\zeta_3]$. In the case of $\mathbb{K} = \mathbb{Q}[\zeta_3]$, we need $n + 1$ ramified primes in order to construct a field with class group of 3-rank at least n.

The general case is considered in the following corollary.

Corollary 1 *Let $\mathbb{K}$ be a Galois CM extension and G be a finite abelian group of odd exponent $e = \exp(G)$. Then there is an abelian extension $\mathbb{L}/\mathbb{K}$ of exponent e whose class group contains a copy of G.*

Proof When $e = p^j$ is a prime power, we argue like in Lemma 2, the proof of which does not depend on the exponent j being equal to 1, since the Hasse Norm Principle holds in arbitrary cyclic extensions. If e is composite, one argues separately for each prime power dividing e and uses the fact that the abelian group G is a direct sum of its p-Sylow subgroups. □

Acknowledgements F. L. was supported in part by grant CPRR160325161141 and an A-rated scientist award both from the NRF of South Africa and by grant no. 17-02804S of the Czech Granting Agency. Part of this work was done during a visit of F. L. at the Max Planck Institute for Mathematics in Bonn in March 2017. He thanks this Institute for hospitality and support.

References

1. Yu. F. Bilu, F. Luca, Divisibility of class numbers: enumerative approach. J. Reine Angew. Math. **578**, 79–91 (2005)
2. Y. Bilu, J. Gillibert, Chevalley–Weil theorem and subgroups of class groups, arXiv:1606.03128
3. R. Cramer, L. Ducas, B. Wesolowski, Short Stickelberger class relations and application to ideal–SVP, in *Advances in Cryptology* (EUROCRYPT, 2017), pp. 324–348
4. A. Levin, Ideal class groups, Hilbert's irreducibility theorem, and integral points of bounded degree on curve. J. Th. Nombres Bordeaux **19**, 485–499 (2007)
5. J.-F. Mestre, Corps quadratiques dont le 5-rang du groupe des classes est ≥ 3. C. R. Acad. Sci. Paris (Série 1) **315**, 371–374 (1992)
6. L. Washington, *Introduction to Cyclotomic Fields (Volume 83 of Graduate Texts Mathematics)*, 2nd edn. (Springer, 1996)

Cyclotomic Numbers and Jacobi Sums: A Survey

Md. Helal Ahmed and Jagmohan Tanti

2010 Mathematics Subject Classification Primary: 11T24 · Secondary: 11T22

1 Introduction

The Jacobi's tremendous mathematical legacy has many contributions to the field of mathematics, among which are the Jacobi symbol, the Jacobi triple product, the Jacobian in the change of Variables theorem and the Jacobi elliptic functions. Among his multiple discoveries, Jacobi sums appear as one of the most important findings. In any given finite field $\mathbb{F}_q$, Jacobi sums of order e mainly depend on two parameters. Therefore, these values could be naturally assembled into a matrix of order e. Jacobi initially proposed these sums as mathematical objects, and for more certainty, he mailed them to Gauss in 1827 (see [11, 29]). After 10 years, Jacobi [30] published his findings including all the extensions provided by other scholars such as Cauchy, Gauss and Eisenstein. It is worth mentioning that while Gauss sums suffice for a proof of quadratic reciprocity, a demonstration of cubic reciprocity law along similar lines requires a foray into the realm of Jacobi sums. In order to prove biquadratic reciprocity, Eisenstein [18] formulated a generalization of Jacobi sums. As illustrated in [27], Jacobi sums could be used for estimating the number of integral solutions to congruences such as $x^3 + y^3 \equiv 1 \pmod{p}$. These estimates played a key role in the development of Weil conjectures [50]. Jacobi sums could be used for the determination of a number of solutions of diagonal equations over finite fields. Jacobi

Md. Helal Ahmed (✉) · J. Tanti
Department of Mathematics, Central University of Jharkhand,
Ranchi 835205, India
e-mail: ahmed.helal@cuj.ac.in; ahmedhelal91@gmail.com

J. Tanti
e-mail: jagmohan.t@gmail.com

K. Chakraborty et al. (eds.), *Class Groups of Number Fields and Related Topics*,
https://doi.org/10.1007/978-981-15-1514-9_12

sums were also utilized in primality test by Adleman, Pomerance, and Rumely [2]. The Problem of congruences of Jacobi sums of order e concerns to establish certain congruence conditions modulo an appropriate element, which is useful to determine an element in $\mathbb{Z}[\zeta_e]$ as a Jacobi sum of order e. It is worth mentioning that congruence conditions for Jacobi sums play major role for determination of algebraic characterization/ diophantine systems of Jacobi sums, hence of all Jacobi sums together with the absolute value and prime ideal decomposition of Jacobi sums.

Cyclotomic number is one of the most important objects in number theory and in other branches of mathematics. These number have been extensively used in coding theory, cryptography, and in other branches of information theory. One of the central problems in the study of these numbers is the determination of all cyclotomic numbers of a specific order for a given field in terms of solutions of certain Diophantine system. This problem has been treated by many mathematicians including Gauss who had determined all the cyclotomic numbers of order 3 in the field $\mathbb{F}_q$ with prime $q \equiv 1 \pmod 3$. Complete solutions to this cyclotomic number problem have been computed for some specific orders. For instance, the cyclotomic numbers of prime order e in the finite field $\mathbb{F}_q$ with $q = p^r$ and $p \equiv 1 \pmod e$ have been investigated by many authors (see [32] and the references therein). Cyclotomic numbers of order e over the field $\mathbb{F}_q$ with characteristic p, in general, cannot be determined only in terms of p and e, but that one requires a quadratic partition of q too.

In this survey article, we discuss some interesting results concerning the Jacobi sums and its congruences, and cyclotomic numbers as well as the current status of the problem. Starting from Gauss, this topic has been studied extensively by many authors and thus there exist a large number of research articles. Due to the versatility, this survey may miss out some interesting references and thus, some interesting results too and thus this article is never claimed to be a complete one.

2 Definitions and Notations

Let $e \geq 2$ be an integer, p a rational prime, $q = p^r$, $r \in \mathbb{Z}^+$ and $q \equiv 1 \pmod e$. Let $\mathbb{F}_q$ be a finite field of q elements. We can write $q = p^r = ek + 1$ for some $k \in \mathbb{Z}^+$. Let γ be a generator of the cyclic group $\mathbb{F}_q^*$ and $\zeta_e = exp(2\pi i/e)$. Also for $a \in \mathbb{F}_q^*$, $\text{ind}_\gamma(a)$ is defined to be a positive integer $m \leq q-1$ such that $a = \gamma^m$. Define a multiplicative character $\chi_e : \mathbb{F}_q^* \longrightarrow \mathbb{Q}(\zeta_e)$ by $\chi_e(\gamma) = \zeta_e$ and extend it on $\mathbb{F}_q$ by putting $\chi_e(0) = 0$. For integers $0 \leq i, j \leq e-1$, the Jacobi sum $J_e(i, j)$ is defined by

$$J_e(i, j) = \sum_{v \in \mathbb{F}_q} \chi_e^i(v) \chi_e^j(v+1).$$

However, in the literature a variation of Jacobi sums are also considered and is defined by

$$J_e(\chi_e^i, \chi_e^j) = \sum_{v \in \mathbb{F}_q} \chi_e^i(v)\chi_e^j(1-v).$$

Observe that $J_e(i, j) = \chi_e^i(-1)J_e(\chi_e^i, \chi_e^j)$. When $q = 2^r$, $\chi_e^i(-1) = \chi_e^i(1) = 1$ and both the Jacobi sums coincide. Otherwise $\chi_e^i(-1) = (-1)^{ik}$ and hence the two Jacobi sums differ at most by sign.

For $0 \le a, b \le e-1$, the cyclotomic number $(a, b)_e$ of order e is defined as the number of solutions (s, t) of the following:

$$\gamma^{es+a} + \gamma^{et+b} + 1 \equiv 0 \pmod{q}; \qquad 0 \le s, t \le k-1. \tag{2.1}$$

or

One can define, for $0 \le a, b \le e-1$, the cyclotomic numbers $(a, b)_e$ of order e is as follows:

$$\begin{aligned}(a, b)_e &:= \#\{v \in \mathbb{F}_q \mid \chi_e(v) = \zeta_e^a,\ \chi_e(v+1) = \zeta_e^b\} \\ &= \#\{v \in \mathbb{F}_q \setminus \{0, -1\} \mid \text{ind}_\gamma v \equiv a \pmod{e},\ \text{ind}_\gamma(v+1) \equiv b \pmod{e}\}.\end{aligned}$$

The cyclotomic numbers $(a, b)_e$ and the Jacobi sums $J_e(i, j)$ are well connected by the following relations [11, 46]:

$$\sum_a \sum_b (a, b)_e \zeta_e^{ai+bj} = J_e(i, j), \tag{2.2}$$

and

$$\sum_i \sum_j \zeta_e^{-(ai+bj)} J_e(i, j) = e^2 (a, b)_e. \tag{2.3}$$

(2.2) and (2.3) together show that if we want to calculate all the cyclotomic numbers $(a, b)_e$ of order e, it is sufficient to calculate all the Jacobi sums $J_e(i, j)$ of the same order, and vice-versa.

3 Properties of Jacobi Sums and Cyclotomic Numbers

We begin this section with the following theorem which is recalled from [1, 4, 11].

Theorem 3.1 *The following statements hold*

(i) *If* $m + n + s \equiv 0 \pmod{e})$ *then*

$$\begin{aligned}J_e(m, n) = J_e(s, n) &= \chi_e^s(-1)J_e(s, m) \\ &= \chi_e^s(-1)J_e(n, m)\end{aligned}$$

$$= \chi_e^m(-1)J_e(m,s)$$
$$= \chi_e^m(-1)J_e(n,s).$$

In particular,

$$J_e(1,m) = \chi_e(-1)J_e(1,s) = \chi_e(-1)J_e(1,e-m-1).$$

(ii) $J_e(0,j) = \begin{cases} -1 & if\ j \not\equiv 0\ (mod\ e), \\ q-2 & if\ j \equiv 0\ (mod\ e). \end{cases}$

(iii) $J_e(i,0) = -\chi_e^i(-1)\ if\ i \not\equiv 0\ (mod\ e)$.

(iv) *If* $m+n \equiv 0\ (mod\ e)$ *but not both* m *and* n *zero modulo* e, *then* $J_e(m,n) = -1$.

(v) *For* $(k,e) = 1$ *and* σ_k *a* $\mathbb{Q}$ *automorphism of* $\mathbb{Q}(\zeta_e)$ *with* $\sigma_k(\zeta_e) = \zeta_e^k$, *we have* $\sigma_k J_e(m,n) = J_e(mk,nk)$. *In particular, if* $(m,e) = 1$, m^{-1} *denotes the inverse of* $m \pmod e$ *then* $\sigma_{m^{-1}} J_e(m,n) = J_e(1,nm^{-1})$.

(vi) $J_{2e}(2m,2s) = J_e(m,n)$.

(vii) $J_e(1,n)\overline{J_e(1,n)} = \begin{cases} q & if\ n \not\equiv 0,-1\ (mod\ e), \\ 1 & if\ n \equiv 0,-1\ (mod\ e). \end{cases}$

(viii) *Let* m, n, s *be integers and* l *be an odd prime, such that* $m+n \not\equiv 0\ (mod\ 2l)$ *and* $m+s \not\equiv 0\ (mod\ 2l)$. *Then*

$$J_{2l}(m,n)J_{2l}(m+n,s) = \chi^m(-1)J_{2l}(m,s)J_{2l}(n,s+m).$$

(ix) *Let* m, n, s *be integers and* l *be an odd prime, such that* $m+n \not\equiv 0\ (mod\ 2l^2)$ *and* $m+s \not\equiv 0\ (mod\ 2l^2)$. *Then*

$$J_{2l^2}(m,n)J_{2l^2}(m+n,s) = \chi^m(-1)J_{2l^2}(m,s)J_{2l^2}(n,s+m).$$

In the next theorem, we state some basic properties of the cyclotomic numbers of order e.

Theorem 3.2 ([11], Berndt) *The cyclotomic numbers of order e have the following properties:*

i. $(a,b)_e = (a',b')_e$ *if* $a \equiv a' \pmod e$ *and* $b \equiv b' \pmod e$.

ii. $(a,b)_e = (e-a,b-a)_e$ *along with the following:*

$$(a,b)_e = \begin{cases} (b,a)_e & \text{if } k \text{ is even or } q = 2^r, \\ (b+\frac{e}{2}, a+\frac{e}{2})_e & \text{otherwise.} \end{cases}$$

iii.

$$\sum_{a=0}^{e-1}\sum_{b=0}^{e-1}(a,b)_e = q-2,$$

iv.

$$\sum_{b=0}^{e-1}(a,b)_e = k - n_a,$$

where n_a is given by

$$n_a = \begin{cases} 1 & \text{if } a = 0, 2 \mid k \text{ or if } a = \frac{e}{2}, 2 \nmid k; \\ 0 & \text{otherwise.} \end{cases}$$

v.

$$\sum_{a=0}^{e-1}(a,b)_e = \begin{cases} k-1 & \text{if } b = 0; \\ k & \text{if } 1 \le b \le e-1. \end{cases} \tag{3.1}$$

vi. $(a,b)'_e = (r^0 a, r^0 b)_e$,
where the prime ($'$) indicates that the cyclotomic number is taken with respect to the generator γ^{r_0} in place of γ in $\mathbb{F}_q^*$.

4 Jacobi Sums and It Congruences

Gauss theories represented the cornerstone of Jacobi sum findings. Many research work have been conducted by a number of mathematicians in an attempt to find out the Diophantine system that characterize the coefficients of Jacobi sums, i.e., giving a Diophantine system whose unique solution provides the coefficients of a particular Jacobi sum. Jacobi sums are particularly used for obtaining the cyclotomic numbers of the same order and vice-versa (i.e., the cyclotomic numbers of order e are known if one knows all the Jacobi sums of order e and vice-versa). Evaluating all the Jacobi sums of order e is relatively intricate. A number of authors devoted for the evaluation of Jacobi sums with certain order. Obtaining the concerned relations helps in reducing the complexity of evaluating all Jacobi sums as well as the cyclotomic numbers. The evaluations and relationships of Jacobi sums of orders 3, 4, and 7 were introduced by Jacobi himself in a letter [29] to Gauss in 1827. Relationship between the sums of orders e for $e \le 6$, $e = 8$, 10, and 12 were established by Dickson [15]. In later stages, Muskat [41] established the relation of order 12 in terms of the fourth root of unity to resolve the sign of ambiguity. Dickson [17] found specific relationships for sums of orders 15, 16, 20, and 24. Muskat [41] developed Dicksons work for $e = 15$ and 24 and extended it to sums of order 30. Complete methods of $e = 16$ and 20 exist in Whiteman [53] and Muskat [42], respectively. In fact, before Dickson's work [15–17], Western [51] determined Jacobi sums of orders 8, 9, and 16. An important issue that should be borne in mind is that all theories showed that Jacobi sums of higher orders can be expressed in terms of Jacobi sums of lower orders. Dickson [16] gave some particular relationships for sums of orders 14 and 22. Muskat [40] provided complete results for order 14. Dickson [17] also investigated sums of orders 9 and 18, while Baumert and Fredrickson [8] gave corrections to some of his results and removed the sign of ambiguity. Zee also found relationships for sums of orders 13 and 60 in [57], and investigated the sums of order 22 in [58]. Relationships for orders 21, 28, 39, 55, and 56 are provided in one of Muskat and Zee research works [43]. Berndt and Evans [9] obtained sums of orders 3, 4, 6, 8, 12, 20, and 24 and they also determined sums of orders 5, 10, and 16 in [10].

Parnami et al. [45] showed that for an odd prime l, it is sufficient to calculate $J_l(1,(l-3)/2)$ number of Jacobi sums for $l>3$ and $J_l(1,1)$ for $l=3$ to obtain all the Jacobi sums of order l. Thus, it reduced the complexity to $l^2-(l-3)/2$ for $l>3$ and l^2-1 for $l=3$. Acharya and Katre [1] indicated that calculating all the Jacobi sums of order $2l$ is not essential, and it is enough to calculate $J_{2l}(1,n)$ for $1 \leq n \leq 2l-3$, n odd or $1 \leq n \leq 2l-2$, n even number of Jacobi sums. In [3], we showed that Jacobi sums of order $2l^2$ can be determined from the Jacobi sums of order l^2. The Jacobi sums of order $2l^2$ can also be obtained from $J_{2l^2}(1,n)$, $1 \leq n \leq 2l^2-3$ for n odd (or equivalently, $2 \leq n \leq 2l^2-2$ for n even). Further the Jacobi sums of order l^2 can be evaluated if one knows the Jacobi sums $J_{l^2}(1,i)$, $1 \leq i \leq \frac{l^2-3}{2}$.

For some small values of e the study of congruences of Jacobi sums is available in the literature. For l an odd prime, Dickson [16] obtained the congruences $J_l(1,n) \equiv -1 \pmod{(1-\zeta_l)^2}$ for $1 \leq n \leq l-1$. Parnami, Agrawal and Rajwade [45] also calculated this separately. Iwasawa [28] in 1975, and in 1981 Parnami, Agrawal, and Rajwade [44] showed that the above congruences also hold $\pmod{(1-\zeta_l)^3}$. Further in 1995, Acharya and Katre [1] extended the work on finding the congruences for Jacobi sums and showed that

$$J_{2l}(1,n) \equiv -\zeta_l^{m(n+1)} \pmod{(1-\zeta_l)^2},$$

where n is an odd integer such that $1 \leq n \leq 2l-3$ and $m = \text{ind}_\gamma 2$. Also in 1983, Katre and Rajwade [31] obtained the congruence of Jacobi sum of order 9, i.e.,

$$J_9(1,1) \equiv -1-(\text{ind } 3)(1-\omega) \pmod{(1-\zeta_9)^4},$$

where $\omega = \zeta_9^3$. In 1986, Ihara [26] showed that if $k>3$ is an odd prime power, then

$$J_k(i,j) \equiv -1 \pmod{(1-\zeta_k)^3}.$$

Evans ([21], 1998) used simple methods to generalize this result for all $k>2$, getting sharper congruences in some cases, especially when $k>8$ is a power of 2. Congruences for the Jacobi sums of order l^2 ($l>3$ prime) were obtained by Shirolkar and Katre [46]. They showed that

$$J_{l^2}(1,n) \equiv \begin{cases} -1+\sum_{i=3}^{l} c_{i,n}(\zeta_{l^2}-1)^i \pmod{(1-\zeta_{l^2})^{l+1}} & if \ \gcd(l,n)=1, \\ -1 \pmod{(1-\zeta_{l^2})^{l+1}} & if \ \gcd(l,n)=l. \end{cases}$$

Recently, we [3] have determined the congruences $\pmod{(1-\zeta_{l^2})^{l+1}}$ for Jacobi sums of order $2l^2$ in terms of the coefficients of Jacobi sums of order l. We split the problem into two cases:
Case 1. n is odd. This case splits into four subcases:
Subcase i. $n = l^2$.
Subcase ii. $n = dl$, where $1 \leq d \leq 2l-1$, d is an odd and $d \neq l$.
Subcase iii. $1 \leq n < 2l^2 - 1$ with $\gcd(n, 2l^2) = 1$.
Subcase iv. $n = 2l^2 - 1$.
Case 2. n is even. In this case the Jacobi sums $J_{2l^2}(1, n)$ can be calculated using the relation $J_{2l^2}(1, n) = \chi_{2l^2}(-1)J_{2l^2}(1, 2l^2 - n - 1)$. More precisely, we proved

Theorem 4.1 *Let $l \geq 3$ be a prime and $q = p^r \equiv 1 \pmod{2l^2}$. If $1 \leq n \leq 2l^2 - 1$ and $1 \leq d \leq 2l - 1$ are odd integer, then a congruence for $J_{2l^2}(1, n)$ over $\mathbb{F}_q$ is given by*

$$J_{2l^2}(1,n) \equiv \begin{cases} \zeta_{l^2}^{-w}(-1+\sum_{i=3}^{l} c_{i,(l^2-1)/2}(\zeta_{l^2}-1)^i) \pmod{(1-\zeta_{l^2})^{l+1}}, \ \textit{if } n = l^2. \\ -\zeta_{l^2}^{-w(dl+1)}(-1+\sum_{i=3}^{l} c_{i,(l^2-1)/2}(\zeta_{l^2}-1)^i)(-1+\sum_{i=3}^{l} c_{i,dl-1}(\zeta_{l^2}^{(-1-dl)/2}-1)^i) \\ \quad \pmod{(1-\zeta_{l^2})^{l+1}} \ \textit{if } d \neq l \textit{ odd integer and } n = dl. \\ \zeta_{l^2}^{-w(n+1)}(-1+\sum_{i=3}^{l} c_{i,(l^2-1)/2}(\zeta_{l^2}-1)^i)(-1+\sum_{i=3}^{l} c_{i,(l^2-1)/2}(\zeta_{l^2}^{n}-1)^i) \\ (-1+\sum_{i=3}^{l} c_{i,(-1-n)}(\zeta_{l^2}^{(1-l^2)/2}-1)^i) \pmod{(1-\zeta_{l^2})^{l+1}}, \ \textit{if } gcd(n, 2l^2) = 1, 1 \leq n < 2l^2-1. \\ -1 \pmod{(1-\zeta_{l^2})^{l+1}}, \ \textit{if } n = 2l^2-1. \end{cases}$$

where $c_{i,n}$ are as described in the Theorem 4.2 and $w = ind_\gamma 2$ with γ a generator of $\mathbb{F}_q^*$.
If n is even, $2 \leq n \leq 2l^2 - 2$ the congruences for Jacobi sums $J_{2l^2}(1, n)$ can be calculated using the relation $J_{2l^2}(1, n) = \chi_{2l^2}(-1)J_{2l^2}(1, 2l_2 - n - 1)$. Also if d in the theorem is even then $J_{2l^2}(1, dl) = \chi(-1)J_{2l^2}(1, 2l^2 - dl - 1)$ and $2l^2 - dl - 1$ is odd. Thus the congruences for $J_{2l^2}(1, n)$ gets completely determined and hence that of all Jacobi sums of order $2l^2$.

Also in [3], calculated the congruences of Jacobi sums $J_9(1, n)$, $1 \leq n \leq 8$ which is not covered in [46] and revised the result congruences of Jacobi sums of order l^2 for $l \geq 3$ a prime. Hence Theorem 5.4 [46] is precisely revised as the following theorem:

Theorem 4.2 *Let $l \geq 3$ be a prime and $p^r = q \equiv 1 \pmod{l^2}$. If $1 \leq n \leq l^2 - 1$, then a congruence for $J_{l^2}(1, n)$ for a finite field $\mathbb{F}_q$ is given by*

$$J_{l^2}(1,n) \equiv \begin{cases} -1 + \sum_{i=3}^{l} c_{i,n}(\zeta_{l^2} - 1)^i \pmod{(1-\zeta_{l^2})^{l+1}} & if \ \gcd(l,n) = 1, \\ -1 \pmod{(1-\zeta_{l^2})^{l+1}} & if \ \gcd(l,n) = l, \end{cases}$$

where for $3 \leq i \leq l - 1$, $c_{i,n}$ are described by equation (5.3) and $c_{l,n} = S(n)$ is given by Lemma 5.3 in [46].

5 Cyclotomic Numbers

Since the time of Gauss, many authors have approached the problem of determining cyclotomic numbers in terms of the solutions of certain Diophantine systems. Such a problem arises when Gauss solving his period equation in the case in terms of the uniquely determined L of the Diophantine system $4p = L^2 + 27M^2$, $L \equiv 1 \pmod 3$. In 1935, a series of three papers were published by Dickson, in which he reviewed and extended the theory of cyclotomy. In the first one [15], he considered the cases for cyclotomic numbers of order $e \leq 6$, $e = 8$, 10, and 12. In the second paper [16], he explained the general theory for cyclotomic numbers of prime order and twice of a prime order and he clearly studied the cases $e = 14$ and 22. In the third paper [17], he discussed cyclotomic numbers of orders $e = 9$ & 18, and in the last part of this paper, he singled out a part entitles

$$\text{Theory for } \phi(e) = 8,\ e = 15, 16, 20, 24, 30.$$

However, the case $e = 30$ was completely ignored and one relation associated with $e = 16$ was also omitted. The case $e = 15$ was left with the sign of ambiguity and only introductory discussions were given to $e = 20$ and 24. The sign of ambiguity for $e = 15$ was resolved by Muskat [41]. Here he also provided a complete analysis for $e = 24$ and 30.

The cyclotomic number may be defined for $e = 1$; in that case we have $(0, 0)_1 = q - 2$. The determination of cyclotomic numbers of order $e = 2$ in $\mathbb{F}_p$ was considered by Dickson in [15] in terms of the period equation [[15], Sect. 9]. He showed that $(a, b)_2$'s are uniquely determined by $p = ek + 1$ and period equation becomes $\eta^2 + \eta + c = 0$, where $c = -1/4(p - 1)$, if k is even and $c = 1/4(p + 1)$, if k is odd. Again the case $e = 2$ was considered by Whiteman [52] in terms of the Jacobsthal sums by which his Diophantine system become $p = a^2 + b^2$; $a = (\psi_4(1) + 2)/2$ and $b = \psi_4(\gamma)/2$. The delightful book by Davenport [14], for $e = 2$, the cyclotomic numbers do not depend upon the generator γ, they are given by

$$(0, 0)_2 = (q - 5)/4,\ \ (0, 1)_2 = (1, 0)_2 = (1, 1)_2 = (q - 1)/4,\ \ if\ q \equiv 1 \pmod 4;$$

$$(0, 0)_2 = (1, 0)_2 = (1, 1)_2 = (q - 3)/4,\ \ (0, 1)_2 = (q + 1)/4,\ \ if\ q \equiv 3 \pmod 4.$$

The determination of the cyclotomic numbers of order $l = 3$ in $\mathbb{F}_p$ was considered by Gauss in [24] in terms of the solutions of the diophantine system $4p = L^2 + 27M^2$, $L \equiv 1 \pmod 3$, when he obtained his period equation in this case in terms of the uniquely determined L. The three cyclotomic periods of order 3 satisfy $x^3 + x^2 - \dfrac{p-1}{3}x - \frac{1}{27}(3p - 1 + pl) = 0$. These equations determine L uniquely, but M is determined only upto sign. Gauss gave formulae for cyclotomic numbers of order 3 in terms of L and M.

Theorem 5.1 [24] *For a prime* $p \equiv 1 \pmod 3$, *write*

$$4p = L^2 + 27M^2, \ L \equiv 1 \pmod 3.$$

Then the nine cyclotomic numbers of order 3 *are given by*

$$(0,0)_3 = (p - 8 + L)/9,$$

$$(0,1)_3 = (1,0)_3 = (2,2)_3 = (2p - 4 - L + 9M)/18,$$

$$(1,1)_3 = (0,2)_3 = (2,0)_3 = (2p - 4 - L - 9M)/18,$$

$$(1,2)_3 = (2,1)_3 = (p + 1 + L)/9.$$

He says that these formulae give the cyclotomic numbers of order 3 for some generator γ of $\mathbb{F}_{p^*}$. If M is replaced by $-M$ in all the formulae then one gets cyclotomic numbers corresponding to some other generator γ' of $\mathbb{F}_{p^*}$. One says that the cyclotomic problem in $\mathbb{F}_p$ for $l = 3$ was solved by Gauss. However, the solution does not make it clear which sign of M goes with which γ, without an alternative evaluation of some cyclotomic numbers of order 3, say $(1,0)_3$ or $(1,1)_3$. In a footnote to the section 358 of [24] (p. 444, English edition or p. 432, German edition) Gauss remarks: "As far as the ambiguity of the sign of M in $4p = L^2 + 27M^2$, $L \equiv 1 \pmod 3$, for the determination of cyclotomic numbers of order 3, is concerned, it is unnecessary to consider this question here, and by the nature of the case it cannot be determined because it depends on the selection of the primitive root g mod p. For some primitive roots, M will be positive, for others negative". Later, this case $l = 3$ was again taken up by Dickson [15] and considered the same diophantine equation. But his calculation was again a Gauss type of ambiguity. In 1952, Whiteman [52] considered the same case and resolved the sign of ambiguity using Jacobsthal sums and his Diophantine system became $4p = c^2 + 3d^2$; $c \equiv 1 \pmod 3$ and $d \equiv 0 \pmod 3$. Further, Hall [25] and Storer [48] generalized the results of Gauss and Dickson for $l = 3$ to finite fields of $q = p^r$ elements. However, when $p \equiv 1 \pmod 3$, their results for $\mathbb{F}_q$ again have a Gauss and Dickson type of ambiguity.

For a prime p, the theory for cyclotomy [15] is well known. The corresponding theory has been developed for pq in [56]. Further these theories were extended in [47] in connection with the construction of finite difference sets. The above mention cases for p or pq, the cyclotomic constants depend upon one or more representation as binary quadratic forms of the type $s^2 + Dt^2$. In [48], T. Storer established the uniqueness for p^r of cases $e = 3, 4, 6, 8$, by same generalizing procedure in [47]. For the cases $e = 3$ and $e = 4$, the unique proper representation are

$$4p^r = s^2 + 27t^2; \ s \equiv 1 \pmod 3$$

and

$$p^r = s^2 + 4t^2; \quad s \equiv 1 \pmod 4)$$

respectively. Cyclotomic numbers of order e over finite $\mathbb{F}_q$ with characteristic p, in general, can not be determined only in terms of p and e, but that one requires a quadratic partition of q too. The cases for $e = 6$ and $e = 8$, the unique proper representation were established in terms of the binary quadratic form $s^2 + Dt^2$ and two such forms, respectively.

The formulae for cyclotomic numbers of order 4 ($p \equiv 1 \pmod 4$), p prime) in terms of the quadratic partition $p = s_0^2 + t_0^2$, $s_0 \equiv 1 \pmod 4$) was obtained by Gauss [23], which fixes t_0 upto sign and s_0 uniquely. Further, Dickson [15] also worked in this account and his diophantine equation was $p = x^2 + 4y^2$, $x \equiv 1 \pmod 4$. However, Gauss and Dickson did not resolve the sign of ambiguity in t_0 and y respectively, viz., given a generator γ of $\mathbb{F}_{p^*}$, it is not clear that which sign of t_0 and y gives correct formulae for the cyclotomic numbers corresponding to γ. Again the case $e = 4$ was considered by Whiteman [52] and resolve the sign of ambiguity using Jacobsthal sums. The corresponding result of Hall [25] for $\mathbb{F}_q$ by setup $q = p^r \equiv 1 \pmod 4$) also has a similar sign of ambiguity in the case when $q = p \equiv 1 \pmod 4$). Later Katre and Rajwade [34] resolve the sign of ambiguity and they gave the formulae to determine the cyclotomic numbers of order 4 as for k even,

$$\begin{aligned}
(0,0)_4 &= 1/16(q - 11 - 6s), \\
(0,1)_4 &= 1/16(q - 3 + 2s + 4t), \\
(0,2)_4 &= 1/16(q - 3 + 2s), \\
(0,3)_4 &= 1/16(q - 3 + 2s - 4t), \\
(1,2)_4 &= 1/16(q + 1 - 2s),
\end{aligned}$$

and for k odd,

$$\begin{aligned}
(0,0)_4 &= 1/16(q - 7 + 2s), \\
(0,1)_4 &= 1/16(q + 1 + 2s - 4t), \\
(0,2)_4 &= 1/16(q + 1 - 6s), \\
(0,3)_4 &= 1/16(q + 1 + 2s + 4t), \\
(1,1)_4 &= 1/16(q - 3 - 2s).
\end{aligned}$$

The next case $l = 5$ was treated by Dickson [15] for prime p, using the properties of Jacobi sums. Dickson considered the Diophantine system $16p = x^2 + 50u^2 + 50v^2 + 125w^2$, $v^2 - 4uv - u^2 = xw$, and $x \equiv 1 \pmod 5$. These system has exactly eight integral simultaneous solutions. If (x, u, v, w) is one solution, also $(x, -u, -v, w)$ and $(x, \pm u, \mp v, -w)$ are solutions. The remaining four are derived from these four by changing all signs. In terms of these solutions Dickson gave formulae for cyclotomic numbers of order 5. Again, Dickson could not tell which solution goes correctly with γ. Later, Whiteman [52] considered the same diophantine system for cyclotomic numbers of order 5 and resolve the sign of ambiguity using Jacobsthal sums. Katre and Rajwade [33] for the determination of a unique solution, they considered a fifth root of unity in terms of a solution of Dickson's diophantine system.

Determination of cyclotomic numbers of order $e = 6$, Dickson [15] showed that 36 cyclotomic constants $(a, b)_6$ depend solely upon the decomposition $A^2 + 3B^2$ of the prime $p = 6k + 1$. Due to the signs of ambiguity, he took 2 be a cubic residue of p, $\gamma^m \equiv 2 \pmod{p}$ and $m \equiv 1$ *or* $4 \pmod{6}$. The same case again considered by Berndt and Evans [10] by taking $q = p^2$. Later, Hall [25] and Storer [48] extended the case by considering $q = p^r$.

The case $e = 7$ [37], the cyclotomic numbers can be given in terms of Dickson-Hurwitz sums using the work of Muskat [40] or a theorem of Whiteman [54]. In [37], Leonard and Williams obtained the cyclotomic numbers of order $e = 7$ in terms of the solutions of a certain triple of Diophantine equations, analogous to the expressions for the cyclotomic numbers of order 5 in terms of the solutions of a pair of Diophantine equations [54]. They used the result of Muskat [40] to evaluate the cyclotomic numbers of order 7. If $p \equiv 1 \pmod{7}$ then there are exactly six integral simultaneous solutions of the triple of Diophantine equations

$$2x_1^2 + 42(x_2^2 + x_3^2 + x_4^2) + 343(x_5^2 + 3x_6^2) = 72p,$$

$$12x_2^2 - 12x_4^2 + 147x_5^2 - 441x_6^2 + 56x_1x_6 + 24x_2x_3 - 24x_2x_4 + 48x_3x_4 + 98x_5x_6 = 0,$$

$$\begin{aligned}12x_3^2 - 12x_4^2 + 49x_5^2 - 147x_6^2 + 28x_1x_5 + 28x_1x_6 + 48x_2x_3 + 24x_2x_4 + 24x_3x_4 \\ + 490x_5x_6 = 0,\end{aligned}$$

satisfying $x_1 \equiv 1 \pmod{7}$, distinct from the two trivial solutions $(-6t, \pm 2u, \pm 2u, \mp 2u, 0, 0)$, where t is given uniquely and u is given ambiguously by

$$p = t^2 + 7u^2, \qquad t \equiv 1 \pmod{7}.$$

If $(x_1, x_2, x_3, x_4, x_5, x_6)$ is a nontrivial solution with $x_1 \equiv 1 \pmod{7}$ then two others solutions are given by $(x_1, -x_3, x_4, x_2, (-x_5 - 3x_6)/2, (x_5 - x_6)/2)$ and $(x_1, -x_4, x_2, -x_3, (-x_5 + 3x_6)/2, (-x_5 - x_6)/2)$. Each of the other three can be obtained from one given above by changing the signs of x_2, x_3, x_4. The result obtained in [37] is almost similar for $p \equiv 1 \pmod{5}$, result obtained in [15], and which is implicit in the work of Dickson [15, 16], does not appear in the literature.

Dickson [15] showed that in the case of $e = 8$, the 64 cyclotomic constants $(a, b)_8$ depend solely upon the decompositions $p = x^2 + 4y^2$ and $p = a^2 + 2b^2; x \equiv a \equiv 1 \pmod 4$, where the signs of y and b depend on the choice of the generator γ. There are four sets of formulas depending on whether k is even or odd and whether 2 is a biquadratic residue or not. Further, Lehmer [36] improved the results of Dickson and gave the complete table of cyclotomic numbers $(a, b)_8$ of order 8.

The cyclotomic problem for $e = 9$ was studied by Dickson [17] and he gave a simple complete theory for $e = 9$. Each cyclotomic numbers are expressed as a constant plus a linear combination of $p,\ L,\ M,\ c_0,\ c_1,\ c_2,\ c_3,\ c_4,\ c_5$, where

$$4p = L^2 + 27M^2, \quad L \equiv 7 \pmod 9$$

and

$$p = \left(\sum_{i=0}^{5} c_i \beta^i\right)\left(\sum_{i=0}^{5} c_i \beta^{-i}\right) \quad (\beta \text{ be a primitive ninth root of unity})$$

is a factorization of p in the field of ninth roots of unity. Further, he provided a theorem to make the correct choice M. In 1967, again the case $e = 9$ were considered by Baumert and Fredricksen [8]. They carried out the result of Dickson [17]. But they worked little bit further Dickson [17] did. They gave simple relation to choose M as Dickson gave.

The cyclotomic numbers of order 10 was initially considered by Dickson [15], later the same case discussed elaborately by Whiteman [54].

The case $e = 11$ were considered by Leonard and Williams [38]. They considered the Diophantine equations

$$1200p = 12w_1^2 + 33w_2^2 + 55w_3^2 + 110w_4^2 + 330w_5^2 + 660(w_6^2 + w_7^2 + w_8^2 + w_9^2 + w_{10}^2),$$

$$\begin{aligned} 0 = {} & 45w_2^2 + 5w_3^2 + 20w_4^2 - 540w_5^2 + 720w_6^2 - 720w_{10}^2 - 288w_1w_5 + 30w_2w_3 \\ & - 120w_2w_4 - 72w_2w_5 + 200w_3w_4 - 360w_3w_5 + 360w_4w_5 + 1440w_6w_7 \\ & - 1440w_6w_8 + 1440w_7w_8 - 1440w_7w_9 + 1440w_8w_9 - 1440w_8w_{10} + 2880w_9w_{10}, \end{aligned}$$

$$\begin{aligned} 0 = {} & 45w_2^2 - 35w_3^2 - 80w_4^2 + 720w_9^2 - 720w_{10}^2 - 144w_1w_4 - 144w_1w_5 \\ & + 150w_2w_3 - 96w_2w_4 - 216w_2w_5 + 160w_3w_4 + 120w_3w_5 + 240w_4w_5 \\ & + 2880w_6w_7 - 1440w_6w_9 + 1440w_7w_8 - 1440w_7w_{10} + 1440w_8w_9 \\ & + 1440w_8w_{10} + 1440w_9w_{10}, \end{aligned}$$

$$
\begin{aligned}
0 = {} & 45w_2^2 + 5w_3^2 + 20w_4^2 - 540w_5^2 + 720w_7^2 - 720w_{10}^2 - 96w_1w_3 - 48w_1w_4 \\
& - 144w_1w_5 + 126w_2w_3 + 108w_2w_4 - 360w_2w_5 + 20w_3w_4 - 60w_3w_5 \\
& + 600w_4w_5 + 1440w_6w_7 + 1440w_6w_8 - 1440w_6w_{10} + 1440w_7w_8 \\
& + 1440w_7w_{10} + 1440w_9w_{10} + 2880w_8w_9,
\end{aligned}
$$

$$
\begin{aligned}
0 = {} & 27w_2^2 + 35w_3^2 - 40w_4^2 - 360w_5^2 + 720w_8^2 - 720w_{10}^2 - 72w_1w_2 - 24w_1w_3 \\
& - 48w_1w_4 - 144w_1w_5 + 114w_2w_3 + 48w_2w_4 + 144w_2w_5 + 320w_3w_4 \\
& + 1440w_6w_7 + 1440w_6w_9 + 1440w_6w_{10} + 2880w_7w_8 + 1440w_7w_9 \\
& + 1440w_8w_9 + 1440w_9w_{10},
\end{aligned}
$$

$$w_3 + 2w_4 + 2w_5 \equiv 0 \pmod{11},$$

$$w_2 - w_4 + 3w_5 \equiv 0 \pmod{11}.$$

In terms of the solutions of above Diophantine equations, they gave the complete formulae for cyclotomic numbers of order 11. Further, determination of cyclotomic problem is somewhat incomplete, when they found the solutions of abovementioned Diophantine systems using Jacobsthal-Whiteman sums.

The cycloyomic problems for $e = 12$ was considered by Dickson [15] and he showed that cyclotomic constants $(a, b)_{12}$ depend solely upon the decomposition $p = x^2 + 4y^2$ and $p = A^2 + 3B^2$ of the prime $p = 12k + 1$, where $x \equiv 1 \pmod 4$ and $A \equiv 1 \pmod 6$. But his analysis depends upon elaborate computations and is not entirely definitive. For settlement of signs of ambiguity for case $e = 12$, he considered lots of small cases, viz., 2 be a cubic residue of p, 3 be a biquadratic residue and non-residue of p. For odd case, again he considered $2m \equiv 2 \pmod{12}$ and $2m \equiv 10 \pmod{12}$. For even case $2m \equiv 8 \pmod{12}$ and $2m \equiv 4 \pmod{12}$. Later, the same case was considered by Whiteman [55] in a different direction. To evaluate the complete solution of cyclotomic, he divided the prime into 12 different classes and obtained formulae holding for different classes.

The study of cyclotomic numbers of order 14 started by Dickson [16] in 1935. Dickson proved that it is possible to represent the cyclotomic numbers $(a, b)_e$ as a linear combination of the Dickson-Hurwitz sums $B_e(i, v)$, $0 \le i \le e - 1$, if e is an odd prime or twice of an odd prime. Dickson's result that given the septic character of 2, the $B_{14}(i, v)$ can be given as a linear combination of the $B_7(i, v)$, has been employed to express the $(a, b)_{14}$ in terms of the $B_7(i, v)$. Muskat [40] determined the cyclotomic numbers $(a, b)_{14}$ explicitly in terms of the $B_e(i, v)$ where e is twice of an odd prime and the transformation is due to the Whiteman [54].

The study of cyclotomic numbers of order 15 was began by Dickson [17] in 1935 and completed by Muskat [41] in 1968. Dickson calculation had a sign of ambiguity. Muskat [41] resolve the sign of ambiguity. In 1986, Frisen, Muskat, Spearman, and Williams [22] considered the same case by setup $q = p^2 \equiv 1 \pmod{15}$. They showed that as $p \equiv 4 \pmod{15}$, 28 different cyclotomic numbers were evaluated by

$$p = A^2 - AB + B^2, \quad A \equiv -1 \pmod 3, \quad B \equiv 0 \pmod 3,$$
$$p = T^2 + 15U^2, \quad T \equiv -1 \pmod 3,$$

and $p \equiv 11 \pmod{15}$, 29 different cyclotomic numbers were evaluated by

$$p = X^2 + 5U^2 + 5V^2 + 5W^2, \quad X \equiv -1 \pmod 5,$$
$$XW = V^2 - UV - U^2.$$

In [13], Buck, Smith, Spearman, and Williams used Dickson and Muskat evaluations of the Jacobi sums of order 15 to obtain the values of the Dickson-Hurwitz sums $B_{15}(i, v)$ of order 15 defined by

$$B_{15}(i, v) = \sum_{h=0}^{14} (h, i - vh)_{15}.$$

Further, they used a special case of a theorem of Friesen, Muskat, Spearman, and Williams [22] and express each cyclotomic number in terms of the Dickson-Hurwitz sums. Using the values of Dickson-Hurwitz sums of order 15, they derived an explicit formulae for the cyclotomic numbers of order 15. Each cyclotomic numbers of order 15 can be expressed as an integral linear combination of the integers p, 1, a, b, c, d, x, u, v, w, b_0 b_1 b_2 b_3 b_4 b_5 b_6 b_7. The integers a, b, c, d, x, u, v, w have the following properties:

$$p = a^2 + 3b^2, \quad a \equiv -1 \pmod 3,$$
$$p = c^2 + 15d^2, \quad c \equiv -1 \pmod 3,$$
$$p = x^2 + 5u^2 + 5v^2 + 5w^2,$$
$$xw = v^2 - uv - u^2, \quad x \equiv -1 \pmod 5.$$

Lehmar [35] raised the question whether or not constants α, β, γ, δ, ϵ can be found such that

$$265(a, b)_{16} = p + \alpha x + \beta y + \gamma a + \delta b + \epsilon, \tag{5.1}$$

at least for some $(a, b)_{16}$ after written the article [36]. To answer this question, she undertook the following experiment on the SWAC (National Bureau of Standards Western Automatic Computer). The cyclotomic constants of order 16 were computed for eight primes p of the form $32n + 1$ for which 2 is not a biquadratic residue. She found that (5.1) is not satisfied for any $(a, b)_{16}$ when the signs of y and b are taken in accordance with the results on cyclotomic constants of order 8. The calculations exhibited eight solutions, while the formula gave only six. A similar computation for primes p of the form $32n + 17$ also led to a negative result. She regretfully conclude that the cyclotomic constants of order 16 are not expressible in terms of these [17] quadratic partitions alone. The SWAC experiment left an open the question that the Eq. (5.1) can be satisfied for any prime p for which 2 is a biquadratic residue. In [53], Whiteman gave formulae for cyclotomic constants of order 16 in affirmative, and six of the cyclotomic constants in terms of parameters of quartic, octic and bioctic Jacobi sums. He further gave a table of formulas for $(a, 0)_{16}$. Later, Evans and Hill [20] gave complete table for cyclotomic numbers of order 16. The computations were performed on the Burroughs 6700 at UCSD by employing the algorithms described in [53]. They stated that it is not possible to accomplish sign resolutions with the use of formulae from Whiteman's result, so they utilized the methodology of [19] to give elementary resolutions of sign ambiguities in quartic and octic Jacobi and Jacobsthal sums in certain cases.

Dickson [17] gave relation to determine the cyclotomic numbers of order 18. But he did not provide a complete table for cyclotomic numbers of order 18. Baumert and Fredricksen [8] considered the case of cyclotomic numbers of order 18 again. They split the solution into cases and introduce the parameters $B = Ind\,2$ and $T = Ind\,3$. Introduced parameters reduced actual parameters to $p,\ L,\ M,\ c_0,\ c_1,\ c_2,\ c_3,\ c_4,\ c_5$ as appeared in the determination of the cyclotomic numbers of order 9. They gave complete listing of all cyclotomic numbers of order 18 (k odd and k even) in an unpublished mathematical tables file of Mathematics of Computation. Further, by the use of cyclotomic formulas given in Table 5 and 6, they proved Theorem 5.2 as an application to difference sets.

Theorem 5.2 *The only residue difference set or modified residue difference set which exists for* $e = 18$ *is the trivial* $19 - 1 - 0$ *difference set.*

In [17], Dickson gave a sketchy discussions for cyclotomic numbers of order $e = 20$. He did not give the exact formulae for cyclotomic numbers of order 20. In 1970, Muskat and Whiteman [42] gave the complete formulae for cyclotomic numbers of order $e = 20$. They obtained the cyclotomic numbers of order 20 in terms of the cyclotomic numbers of orders two & four and Jacobi sums of orders 5, 10 & 20.

In 1982, PAR [45] considered the general e cases; e be an odd prime, $q = p^r$, $p \equiv 1 \pmod{e}$. They indicate a general method for solving the cyclotomic problem over $\mathbb{F}_q$. They calculated the cyclotomic numbers of orders $e \leq 19$. The Diophantine systems they considered is as follows

$$q = \sum_{i=0}^{e-1} a_i^2 - \sum_{i=0}^{e-1} a_i a_{i+1},\ (i.e.,\ 2q = (a_0 - a_1)^2 + (a_1 - a_2)^2 + \cdots + (a_{e-1} - a_0)^2),$$

$$\sum_{i=0}^{e-1} a_i a_{i+1} = \sum_{i=0}^{e-1} a_i a_{i+2} = \cdots = \sum_{i=0}^{e-1} a_i a_{i+(e-1)/2},$$

$$1 + a_0 + a_1 + \cdots + a_{e-1} \equiv 0 \pmod{e},$$

$$a_1 + 2a_2 + 3a_3 + \cdots + (e-1)a_{e-1} \equiv 0 \pmod{e},$$

which generalizes the diophantine systems of Gauss–Dickson and Leonard Williams. Moreover they gave a rejection condition

$$p \nmid \prod_{\lambda((n+1)a)>a} H^{\sigma_a},$$

which fixes certain Jacobi sums upto conjugate. They gave full details of the transformation (for $r = 1$) connecting above mentioned Diophantine system with the classical ones for $l = 3, 5$, but they did not connect the cases for $e > 5$ because the rejection condition was too complicated. But their solutions had again Gauss and Dickson type ambiguity.

Further in 1985, Katre and Rajwade [32] solved the cyclotomic problem for any prime e in $\mathbb{F}_q$, $q = p^r$, $p \equiv 1 \pmod{e}$. For solving the cyclotomic problem, they added a new condition to Parnami et al. [45] diophantine systems, i.e., $p|\overline{H}\prod_{\lambda((n+1)a)>a}(b - \zeta_l^{\sigma_{a^{-1}}})$, where a^{-1} was taken mod l, l be an odd prime and $(a, l) = 1$. For cyclotomic numbers of order 2 in $\mathbb{F}_q$, they told that one can determine the cyclotomic numbers by $(0,0)_2 = (q-5)/4$, $(0,1)_2 = (1,0)_2 = (1,1)_2 = (q-1)/4$ if $q \equiv 1 \pmod 4$, $(0,0)_2 = (1,0)_2 = (1,1)_2 = (q-3)/4$, $(0,1)_2 = (q+1)/4$ if $q \equiv 3 \pmod 4$. The problem arises in Parnami et al. [45] to connect the Jacobi sums and e^{th} root of unity, they considered $\gamma^{(q-1)/e} \equiv 1 \pmod{e}$ gave the proper connection between the Jacobi sums and the eth root of unity mod p. Additional condition in Katre and Rajwade [32] also resolve the sign of ambiguity arose in Parnami et al. [45].

For l an odd prime, Acharya and Katre [1] determined the cyclotomic numbers of order $2l$ over the field $\mathbb{F}_q$ for $q = p^r$ with the prime $p \equiv 1 \pmod{2l}$ in terms of the solutions of the diophantine systems considered for the l case except that the proper choice of the solutions for the $2l$ case was made by additional conditions (iv'), (v'), (vi') which replace the conditions (iv), (v), (vi) used in the l case. These additional conditions determine required unique solutions thereby also giving arithmetic characterization of the relevant Jacobi sums and then the cyclotomic numbers of order $2l$ are determined unambiguously by the following theorem. In the same, they also showed how the cyclotomic numbers of order l and $2l$ can be treated simultaneously.

Theorem 5.3 *[1] Let p and l be odd rational primes, $p \equiv 1 \pmod{l}$ (thus $p \equiv 1 \pmod{2l}$ also), $q = p^r$, $r \geq 1$. Let $q = 2lk + 1$. Let ζ_l and ζ_{2l} be fixed primitive $l - th$ and $2l - th$ roots of unity, respectively. Let ζ_l and ζ_{2l} be related by $\zeta_l = \zeta_{2l}^2$, i.e., $\zeta_{2l} = -\zeta_l^{(l+1)/2}$. Let γ be a generator of $\mathbb{F}_q^*$. Let b be a rational integer such that $b = \gamma^{(q-1)/l}$ in $\mathbb{F}_p$. Let $m = ind_\gamma 2$. Let $J_l(i, j)$ and $J_{2l}(i, j)$ denote the Jacobi sums in $\mathbb{F}_q$ of order l and $2l$ (respectively)related to ζ_l and ζ_{2l} (respectively). For $(m, l) = 1$, let σ_m denote the automorphism $\zeta_l \mapsto \zeta_l^m$ of $\mathbb{Q}(\zeta_l)$. For $(m, 2l) = 1$, let τ_m denote the automorphism $\zeta_{2l} \mapsto \zeta_{2l}^m$ of $\mathbb{Q}(\zeta_l)$. Thus if m is odd then $\sigma_m = \tau_m$ and if m is even then $\sigma_m = \tau_{m+l}$. Let $\lambda(r)$ and $\Lambda(r)$ denote the least nonnegative residues of*

r modulo l and $2l$ (resp.). Let $a_0, a_1, \ldots, a_{l-1} \in \mathbb{Z}$ and let $H = \sum_{i=0}^{l-1} a_i \zeta_l^i$. Consider the arithmetic conditions (or diophantine system)

(i) $q = \sum_{i=0}^{l-1} a_i^2 - \sum_{i=0}^{l-1} a_i a_{i+1}$,
(ii) $\sum_{i=0}^{l-1} a_i a_{i+1} = \sum_{i=0}^{l-1} a_i a_{i+2} = \cdots = \sum_{i=0}^{l-1} a_i a_{i+(l-1)/2}$,
(iii) $1 + a_0 + a_1 + \cdots + a_{l-1} \equiv 0 \pmod{l}$.

Let $1 \le n \le l-2$. If $a_0, a_1, \ldots, a_{l-1}$ satisfy $(i)-(iii)$ together with the additional conditions
(iv) $a_1 + 2a_2 + 3a_3 + \cdots + (l-1)a_{l-1} \equiv 0 \pmod{l}$,
(v) $p \nmid \prod_{\lambda((n+1)m)>m} H^{\sigma_m}$,
(vi) $p | \overline{H} \prod_{\lambda((n+1)m)>m} (b - \zeta_l^{\sigma_{m^{-1}}})$, *where m^{-1} is taken* $\pmod{l}$,
then $H = J_l(1, n)$ for this γ and conversely.

Let $1 \le n \le 2l-3$ be an odd integer. If $a_0, a_1, \ldots, a_{l-1}$ satisfy $(i)-(iii)$ together with the additional conditions
$(iv)'$ $a_1 + 2a_2 + 3a_3 + \cdots + (l-1)a_{l-1} \equiv u(n+1) \pmod{l}$,
$(v)'$ $p \nmid \prod_{\Lambda((n+1)m)>m} H^{\tau_m}$,
$(vi)'$ $p | \overline{H} \prod_{\Lambda((n+1)m)>m} (b - \zeta_l^{\tau_{m^{-1}}})$, *where m^{-1} is taken* $\pmod{2l}$,
then $H = J_{2l}(1, n)$ for this γ and conversely.

(In $(v)'$ and $(vi)'$, m varies over only those values which satisfy $1 \le m \le 2l-1$ and $(m, 2l) = 1$.)

Moreover, for $1 \le n \le l-2$ if $a_0, a_1, \ldots, a_{l-1}$ satisfy the conditions $(i)-(vi)$ and if we fix $a_0 = 0$ at the outset and write the a_i corresponding to a given n as $a_i(n)$ then we have $J_l(1, n) = \sum_{i=1}^{l-1} a_i(n) \zeta_l^i$ and the cyclotomic numbers of order l are given by
$l^2(i, j)_l = q - 3l + 1 + \varepsilon(i) + \varepsilon(j) + \varepsilon(i-j) + l \sum_{n=1}^{l-2} a_{in+j}(n) - \sum_{n=1}^{l-2} \sum_{k=1}^{l-1} a_k(n)$
where

$$\varepsilon(i) = \begin{cases} 0 & if\, l|i, \\ l & otherwise, \end{cases}$$

and the subscripts in $a_{in+j}(n)$ are considered modulo l.

Similarly, for n odd, $1 \le n \le 2l-3$, if $a_0, a_1, \ldots, a_{l-1}$ satisfy the conditions $(i)-(iii)$ and $(iv)'-(vi)'$ and if we fix $a_0 = 0$ at the outset and write the a_i corresponding to a given n as $b_i(n)$ then we have $J_{2l}(1, n) = \sum_{i=1}^{l-1} b_i(n) \zeta_l^i$ and the $4l^2$ cyclotomic numbers $(i, j)_{2l}$ are given by

$$4l^2(i,j)_{2l} = q - 3l + 1 + \varepsilon(i) + \varepsilon(j) + \varepsilon(i-j) + l\sum_{n=1}^{l-2} a_{in+j}(n) - \sum_{n=1}^{l-2}\sum_{k=1}^{l-1} a_k(n)$$

$$- \{(-1)^j + (-1)^{i+k} + (-1)^{i+j}\}\{l + \sum_{k=0}^{l-1} b_k(l) + \sum_{u=0}^{l-2}\sum_{k=0}^{l-1} b_k(2u+1)\}$$

$$+ (-1)^j l\{b_{\nu(-i)}(l) + \sum_{u=0}^{l-1} b_{\nu(j-2iu-2i)}(2u+1)\} + (-1)^{i+j} l\{b_{\nu(j)}(l)$$

$$+ \sum_{u=0}^{l-1} b_{\nu(i+2ju+j)}(2u+1)\} + (-1)^{i+k} l\{b_{\nu(-j)}(l) + \sum_{u=0}^{l-1} b_{\nu(i-2ju-2j)}(2u+1)\},$$

where

$$\nu(j) = \begin{cases} \Lambda(j)/2 & \text{if } j \text{ is even,} \\ \Lambda(j+l)/2 & \text{if } j \text{ is odd,} \end{cases}$$

and $\Lambda(r)$ is defined as the least nonnegative residue of r modulo $2l$.

For $q \equiv 1 \pmod{l}$, but p does not necessary $\equiv 1 \pmod{l}$. Let r be the least positive integer such that $q = p^r \equiv 1 \pmod{l}$. In the case when r is even, Anuradha and Katre [6] obtained the Jacobi sums and cyclotomic numbers of order l and $2l$ just in terms of q. For $e = l,\ 2l$ and when r is odd, the cyclotomic problem was treated by Anuradha [7] in her Ph.D. thesis, and thus the problem for the order e is settled for all $q \equiv 1 \pmod{e}$, for $e = l,\ 2l$. Again, for such primes, Shirolkar and Katre [46] obtained formulae for cyclotomic numbers of order l^2 in terms of the coefficients of Jacobi sums of orders l and l^2. Recently, Ahmed et al. [4] obtained formulae for the determination of cyclotomic numbers of orders $2l^2$ in terms of the cyclotomic numbers of orders l, $2l$, l^2 and the coefficient of some special types of Jacobi sums of order l^2 and $2l^2$.

6 Concluding Remarks

Recently, Ahmed et al. [5] introduced a new idea to the construction of cyclotomic matrix and further, the author's developed a secured Public-key cryptography model applying the principle of cyclotomic matrices. Earlier, Leung et al. [39] constructed Hadamard matrices of order $4p^2$ obtained from Jacobi sums of order 16 and they proved the following result.

Theorem 6.1 *Let $p \equiv 7 \pmod{16}$ be a prime. Then there are integers a, b, c, d with*

$$a \equiv 15 \pmod{16},$$

$$b \equiv 0 \pmod{4},$$

$$q^2 = a^2 + 2(b^2 + c^2 + d^2),$$

$$2ab = c^2 - 2cd - d^2.$$

If

$$q = a \pm 2b \text{ or}$$

$$q = a + \delta_1 b + 4\delta_2 c + 4\delta_1\delta_2 4d \text{ with } \delta_i = \pm 1,$$

then there is a regular Hadamard matrix of order $4q^2$.

Betsumiya, Hirasaka et al. [12] gave upper bounds for cyclotomic numbers of order e over a finite field with q elements, where e is a positive divisor of $q - 1$. In particular, they showed that under certain assumptions, cyclotomic numbers are at most $\left\lceil \frac{k}{2} \right\rceil$, and the cyclotomic number $(0, 0)_e$ is at most $\left\lceil \frac{k}{2} \right\rceil - 1$, where $k = (q - 1)/e$. They proved the following.

Theorem 6.2 *Let q be a power of an odd prime p and k a positive divisor of $q - 1$. Then we have the following:*
(i) $(a, b)_e \leq \left\lceil \frac{k}{2} \right\rceil$ for all a, b with $0 \leq a, b < e$ if $p > \frac{3k}{2} - 1$;
(ii) $(a, b)_e \leq \left\lceil \frac{k}{2} \right\rceil - 1$ for each a with $0 \leq a < e$ if k is odd and $p > \frac{3k}{2}$;
(iii) $(0, 0)_e \leq \left\lceil \frac{k}{2} \right\rceil - 1$ if $p > \frac{3k}{2}$;
(iv) $(0, 0)_e = 2$ if p is sufficiently large compared to k and $6|k$;
(v) $(0, 0)_e = 0$ if p is sufficiently large compared to k and $6 \nmid k$.

P. van Wamelen [49] has characterized the Jacobi sums of order e corresponding to any given generator γ of $\mathbb{F}_q^*$. So far this is the most satisfactory solution of the problem, as it takes up the case when e is any integer ≥ 3 and q is any prime power for which $q \equiv 1 \pmod{e}$ and he proved.

Theorem 6.3 *Let $e \geq 3$, p a prime, $q = p^r \equiv 1 \pmod{e}$. Let $q = ef + 1$. Let $g_m = gcd(e, m)$, $g_n = gcd(e, n)$, $g = gcd(e, m + n)$, $g_0 = gcd(g_m, g_n)$. Let $\epsilon_g(k) = 1$ if $g|k$, and 0 if $g \nmid k$. There is a unique polynomial $H \in \mathbb{Z}[x]$ such that $H(x) = a_0 + a_1x + a_2x^2 + \cdots + a_{e-1}x^{e-1}$ and the coefficients satisfy the following three conditions:*
1. (a)

$$\sum_{j=0}^{e-1} a_j^2 = q + g_0 ef^2 - f(g_m + g_n + g).$$

(b) For $k = 1, 2, \ldots, e - 1$

$$\sum_{j=0}^{e-1} a_j a_{j-k} = \epsilon_{g_0}(k) g_0 ef^2 - \epsilon_{g_m}(k) f g_m - \epsilon_{g_n}(k) f g_n - \epsilon_g(k) f g$$

where we consider the subscripts of the a's modulo e.
2.

$$\sum_{k=0}^{e-1} ka_k \equiv \begin{cases} 0 \pmod{e} & \text{if } e \text{ is odd,} \\ (q-1)/2(g_m+g_n) \pmod{e} & \text{if } e \text{ is even.} \end{cases}$$

3. For every d dividing e let $B_d \in \mathbb{Z}[x]$ *be such that its reduction modulo p is the minimal polynomial of* $\gamma^{(q-1)/d}$ *over* $\mathbb{F}_p$ *and* $\prod_{k\in D_d} B_d(\zeta_d^k)$ *is not divisible by* p^2 *in* $\mathbb{Z}[\zeta_d]$*. Then* $H(\zeta_d)$ *must satisfy the following conditions:*
(a) if none of m, n and $m+n$ *are divisible by d,*

$$q|H(\zeta_d)\prod_{k\in D_d} B_d(\zeta_d^{k^{-1}})^{(s_{d.q}(mk)+s_{d.q}(nk)-s_{d.q}(mk+nk))/(p-1)},$$

where k^{-1} *is taken modulo d.*
(b) if $m \equiv -n \not\equiv 0 \pmod{d}$

$$H(\zeta_d) = -\chi_d^m(-1),$$

(c) if exactly one of m and n are divisible by d

$$H(\zeta_d) = -1,$$

(d) if both m and n are divisible by d

$$H(\zeta_d) = q-2.$$

If H is the unique polynomial satisfying these three conditions, then

$$H(\zeta_d) = J(\chi^m, \chi^n).$$

Acknowledgements The authors acknowledge Central University of Jharkhand, Ranchi, Jharkhand for providing necessary and excellent facilities to carry out this research. M H Ahmed would like to thank K. Chakraborty and A. Hoque for their kind invitation and hospitality at Harish-Chandra Research Institute, where this paper was finalized.

References

1. V.V. Acharya, S.A. Katre, Cyclotomic numbers of orders $2l$, l an odd prime. Acta Arith. **69**(1), 51–74 (1995)
2. L. Adleman, C. Pomerance, R. Rumely, On distinguishing prime numbers from composite numbers. Ann. Math. **117**, 173–206 (1983)
3. M.H. Ahmed, J. Tanti, Complete congruences of Jacobi sums of order $2l^2$ with prime l. (arXiv:1902.10512 [math.NT] (Preprint))

4. M.H. Ahmed, J. Tanti, A. Hoque, Complete solution to cyclotomy of order $2l^2$ with prime l. Ramanujan J. https://doi.org/10.1007/s11139-019-00182-9 (2019)
5. M.H. Ahmed, J. Tanti, K. Chakraborty, S. Pusph, A Public-key cryptosystem using cyclotomic matrices. (arXiv: 1906.06921v1 [cs.CR] (Preprint))
6. N. Anuradha, S.A. Katre, Number of points on the projective curves $aY^l = bX^l + cZ^l$ and $aY^{2l} = bX^{2l} + cZ^{2l}$ defined over finite fields, l an odd prime. J. Number Theory **77**, 288–313 (1999)
7. N. Anuradha, Arithmetic Characterization and Applications of Jacobi Sums of Order l and $2l$. Ph.D. thesis, University of Pune, 2000
8. L.D. Baumert, H. Fredricksen, The cyclotomic numbers of order eighteen with applications to difference sets. Math. Comp. **21**, 204–219 (1967)
9. B.C. Berndt, R.J. Evans, Sums of Gauss, Jacobi and Jacobsthal. J. Number Theory **11**, 349–398 (1979)
10. B.C. Berndt, R.J. Evans, Sums of Gauss, Eisenstein, Jacobi, Jacobsthal and Brewer. Illinois J. Math. **23**, 374–437 (1979)
11. B.C. Berndt, R.J. Evans, K.S. Williams, *Gauss and Jacobi Sums* (Wiley, A Wiley-Interscience Publication, New York, 1998)
12. K. Betsumiya, M. Hirasaka, T. Komatsu, A. Munemasa, Upper bounds on cyclotomic numbers. Linear Alg. Appl. **438**(1), 111–120 (2013)
13. N. Buck, L. Smith, B.K. Spearman, K.S. Williams, The cyclotomic numbers of order fifteen. Math. Comp. **48**, 67–83 (1987)
14. H. Davenport, *The Higher Arithmetic, An Introduction to the Theory of Numbers*, Chapter VIII by J. H. Davenport, 7th edn. (Cambridge University Press, Cambridge, 1999)
15. L.E. Dickson, Cyclotomy, higher congruences, and Waring's problem. Am. J. Math. **57**, 391–424 (1935)
16. L.E. Dickson, Cyclotomy and trinomial congruences. Trans. Am. Soc. **37**, 363–380 (1935)
17. L.E. Dickson, Cyclotomy when e is composite. Trans. Am. Math. Soc. **38**, 187–200 (1935)
18. G. Eisenstein, Einfacher beweis und verallgemeinerung des fundamental-theorems fur die biquadratischen reste, *Mathematische Werke*, Band I (Chelsea, New York, 1975), pp. 223–245
19. R.J. Evans, Resolution of sign ambiguities of Jacobi and Jacobsthal sums. Pacific J. Math. **81**, 71–80 (1979)
20. R.J. Evans, J.R. Hill, The cyclotomic numbers of order sixteen. Math. Comp. **33**, 827–835 (1979)
21. R.J. Evans, Congruences for Jacobi Sums. J. Number Theory **71**, 109–120 (1998)
22. C. Friesen, J.B. Muskat, B.K. Spearman, K.S. Williams, Cyclotomy of order 15 over $GF(p^2)$, $p \equiv 4, 11 \pmod{15}$. Int. J. Math. Math. Sci. **9**, 665–704 (1986)
23. C.F. Gauss, Theoria Residuorum Biquadraticorum. Werke **2**, 67–92 (1876)
24. C.F. Gauss, Disquisitiones Arithmeticae, Section 358
25. M. Hall, Jr., Characters and cyclotomy, (Proc. Symp. Pure Math. 8), 31–43. Am. Math. Soc., Providence, R. I. (1965)
26. Y. Ihara, Profinite braid groups, Galois representations, and complex multiplications. Ann. Math. **123**, 43–106 (1986)
27. K. Ireland, M. Rosen, *A Classical Introduction to Modern Number Theory*, 2nd edn. (Springer, New York, 1990)
28. K. Iwasawa, A note on Jacobi sums, in *Symposia Math*, vol. 15 (Academic Press, London, 1975), pp. 447–459
29. C.G.J. Jacobi, Brief an Gauss, **8** Februar 1827. [CW: vol. 7, pp. 393–400]
30. C.G.J. Jacobi, Uber die Kreistheilung und Ihre Anwendung auf die Zahlentheorie, Monatsber. Konigl. Akad. Wiss. Berlin (1837), 127–136. (Same paper as J. Reine Angew. Math. **30** (1846), 166–182) [CW: vol. 6, pp. 254–274]
31. S.A. Katre, A.R. Rajwade, On the Jacobsthal sum $\phi_9(a)$ and the related sum $\psi_9(a)$. Math. Scand. **53**, 193–202 (1983)
32. S.A. Katre, A.R. Rajwade, Complete solution of the cyclotomic problem in $\mathbb{F}_q$ for any prime modulus l, $q = p^\alpha$, $p \equiv 1 \pmod{l}$. Acta Arith. **45**, 183–199 (1985)

33. S.A. Katre, A.R. Rajwade, Unique determination of cyclotomic numbers of order five. Manuscr. Math. **53**(1–2), 65–75 (1985)
34. S.A. Katre, A.R. Rajwade, Resolution of the sign ambiguity in the determination of the cyclotomic numbers of order 4 and the corresponding Jacobsthal sum. Math. Scand. **60**, 52–62 (1987)
35. E. Lehmer, On the cyclotomic numbers of order sixteen. Canad. J. Math. **6**, 449–454 (1954)
36. E. Lehmer, On the number of solutions of $u^k + D \equiv \pmod{p}$. Pacific J. Math. **5**, 103–118 (1955)
37. P.A. Leonard, K.S. Williams, The cyclotomic numbers of order seven. Proc. Am. Math. Soc. **51**, 295–300 (1975)
38. P.A. Leonard, K.S. Williams, The cyclotomic numbers of order eleven. Acta Arith. **26**, 367–383 (1975)
39. K.H. Leung, S.L. Ma, B. Schmidt, New Hadamard matrices of order $4p^2$ obtained from Jacobi sums of order 16. J. Comb. Theory Series A **113**(5), 822–838 (2006)
40. J.B. Muskat, The cyclotomic numbers of order fourteen. Acta Arith. **11**, 263–279 (1966)
41. J.B. Muskat, On Jacobi sums of certain composite orders. Trans. Am. Math. Soc. **134**, 483–502 (1968)
42. J.B. Muskat, A.L. Whiteman, The cyclotomic numbers of order twenty. Acta Arith. **17**, 185–216 (1970)
43. J.B. Muskat, Y.C. Zee, Sign ambiguities of Jacobi sums. Duke Math. J. **40**, 313–334 (1973)
44. J.C. Parnami, M.K. Agrawal, A.R. Rajwade, A congruence relation between the coefficients of the Jacobi sum. Indian J. Pure Appl. Math. **12**(7), 804–806 (1981)
45. J.C. Parnami, M.K. Agrawal, A.R. Rajwade, Jacobi sums and cyclotomic numbers for a finite field. Acta Arith. **41**, 1–13 (1982)
46. D. Shirolkar, S.A. Katre, Jacobi sums and cyclotomic numbers of order l^2. Acta Arith. **147**, 33–49 (2011)
47. T. Storer, A Family of Difference Sets. Dissertation, University of Southern California, 1964
48. T. Storer, On the unique determination of the cyclotomic numbers for Galois fields and Galois domains. J. Comb. Theory **2**, 296–300 (1967)
49. P. van Wamelen, Jacobi sums over finite fields. Acta Arith. **102**(1), 1–20 (2002)
50. A. Weil, Number of solutions of equations in a finite field. Bull. Am. Math. Soc. **55**, 497–508 (1949)
51. A.E. Western, An extension of Eisenstein's law of reciprocity II. Proc. London Math. Soc. **7**(2), 265–297 (1908)
52. A.L. Whiteman, Cyclotomy and Jacobsthal sums. Am. J. Math. **74**, 89–99 (1952)
53. A.L. Whiteman, The cyclotomic numbers of order sixteen. Trans. Am. Math. Soc. **86**, 401–413 (1957)
54. A.L. Whiteman, The cyclotomic numbers of order ten. Proc. Sympos. Appl. Math. **10**; Am. Math. Soc., Providence, R. I., 95–111 (1960)
55. A.L. Whiteman, The cyclotomic numbers of order twelve. Acta Arith. **6**, 53–76 (1960)
56. A.L. Whiteman, A family of difference sets. Illinois J. Math. **6**, 107–121 (1962)
57. Y.C. Zee, The Jacobi sums of orders thirteen and sixty and related quadratic decompositions. Math. Z. **115**, 259–272 (1970)
58. Y.C. Zee, The Jacobi sums of order twenty-two. Proc. Am. Math. Soc. **28**, 25–31 (1971)

A Pair of Quadratic Fields with Class Number Divisible by 3

Himashree Kalita and Helen K. Saikia

2010 Mathematics Subject Classification Primary: 11R29 · Secondary: 11R11

1 Introduction

The divisibility properties of the class number of quadratic fields help to understand the structures of their (ideal) class groups. The class number or more precisely the class group of quadratic fields is one of the most fundamental and curious objects in algebraic number theory. This topic has been studied extensively by many authors since the time of Gauss and thus, there is a good amount of research work around it. The existence of infinitely many real (resp. imaginary) quadratic fields each with class number divisible by a given integer $n \geq 2$ is well known. In particular, it was Nagell [15] who proved that there are infinitely many imaginary quadratic fields whose class number is divisible by a given integer $n \geq 2$. On the other hand, for real quadratic field case, Yamamoto [20] and Weinbeger [21] independently showed that there are infinitely many real quadratic fields whose class number is divisible by a given integer $n \geq 2$. However, after three decades, Ichimura [10] proved that the assumptions considered by Weinbeger [21] were not so necessary and therefore a slight modification could be enough. In the current years, it becomes more interesting and important to explicitly determine real as well as imaginary quadratic fields whose class number is divisible by a given integer $n \geq 2$. It is also important to study the structure of the class groups of quadratic fields. In this direction, numerous results

H. Kalita
Department of Mathematics, Darrang College, Tezpur 784001, Assam, India
e-mail: himashree.kalita28@gmail.com

H. K. Saikia (✉)
Department of Mathematics, Gauhati University, Guwahati 781014, Assam, India
e-mail: hsaikia@yahoo.com

K. Chakraborty et al. (eds.), *Class Groups of Number Fields and Related Topics*,
https://doi.org/10.1007/978-981-15-1514-9_13

have been offered by many authors (see [2–6, 8, 9, 11–14, 18]). This paper is along the same lines, but it carries more information since it deals with a pair of quadratic fields.

Let $k_d = \mathbb{Q}(\sqrt{d})$ and $k_m = \mathbb{Q}(\sqrt{m})$ be two quadratic fields with discriminants D_d and D_m, respectively. In this paper, we show that there exist infinitely many pairs (k_d, k_m) with each class number divisible by 3 and satisfying the property that either $D_d \mid D_m$ or $D_m \mid D_d$. More precisely, we prove

Theorem 1.1 *Let* $p \equiv \pm 4 \pmod 9$ *be a prime and* $\ell \geq 1$ *an integer. The class numbers of* $\mathbb{Q}(\sqrt{p^{12\ell+2} - 4})$ *and* $\mathbb{Q}\left(\sqrt{\frac{4-p^{12\ell+2}}{3}}\right)$ *are divisible by* 3. *Moreover, there are infinitely many such pair of quadratic fields with class number divisible by* 3.

Remark 1.1 If D and d denote the discriminants of $\mathbb{Q}(\sqrt{p^{12\ell+2} - 4})$ and $\mathbb{Q}\left(\sqrt{\frac{4-p^{12\ell+2}}{3}}\right)$, respectively, then $d \mid D$.

2 Some Useful Results

In this section, we discuss some important results that are needed to prove the main result. For a square-free positive integer d, by $h(d)$ and $\mathfrak{h}(-d)$ we mean the class number of the real quadratic field $\mathbb{Q}(\sqrt{d})$ and the imaginary quadratic field $\mathbb{Q}(\sqrt{-d})$, respectively. We denote the fundamental unit of $\mathbb{Q}(\sqrt{d})$ by

$$\varepsilon_d = \begin{cases} \frac{t+u\sqrt{d}}{2} & \text{if } d \equiv 1 \pmod 4, \\ t + u\sqrt{d} & \text{if } d \equiv 2, 3 \pmod 4, \end{cases}$$

where t and u are known as the coefficients of ε_d. We first recall the following result of Artin, Ankeny and Chowla [1, Proposition 2.4].

Proposition 2.1 *Let* $m \equiv 1 \pmod 3$ *be a square-free positive integer. If* $d = 3m$, *then the following holds:*

$$\mathfrak{h}(-m) \equiv -\frac{u}{t}\, h(d) \pmod 3.$$

We now assume the following trinomial:

$$p(x) = x^n - ax + b \qquad (a, b \in \mathbb{Z}).$$

Let K be the minimal splitting field of $p(x)$, that is,

$$K = \mathbb{Q}(\alpha_1, \alpha_2, \ldots, \alpha_n),$$

where $\alpha_1, \alpha_2, \ldots, \alpha_n$ are the roots of $p(x)$. Then the discriminant of $p(x)$ is defined as

$$D_p = \prod_{i<j}(\alpha_i - \alpha_j)^2.$$

We are now in a position to recall the following result which was proved in [19].

Proposition 2.2 *If n is a prime, $p(x)$ is irreducible over $\mathbb{Q}$ and the Galois group of K over $\mathbb{Q}$ is a symmetric group S_n of n items, then K is unramified extension of $\mathbb{Q}(\sqrt{D_p})$ with Galois group A_n.*

3 Proof of the Theorem 1.1

In this section, we prove Theorem 1.1 using the results that have been discussed in Sect. 2 and Siegel's theorem on integral points on a curve.

We consider the trinomial

$$f(x) = x^3 - 3x + p^{6\ell+1},$$

where ℓ is a positive integer and p is a prime of the form:

$$p \equiv \pm 4 \pmod 9.$$

One gets the discriminant of $f(x)$ as follows:

$$D_f = 27(4 - p^{2(6\ell+1)}).$$

Since $p \equiv \pm 4 \pmod 9$, so that $p^{2(6\ell+1)} - 4 \equiv 3 \pmod 9$. This shows that $3 \mid (p^{2(6\ell+1)} - 4)$, and thus one can write $p^{2(6\ell+1)} - 4 = 3m$ for some positive integer m. Therefore

$$D_f = -81m.$$

We see that $p^{2(6\ell+1)} - 4 \equiv 3 \pmod 9$ and $p^{2(6\ell+1)} - 4 = 3m$ together imply $m \equiv 1 \pmod 3$.

Proposition 3.1 *3 divides the class number of the imaginary quadratic field $\mathbb{Q}(\sqrt{-m})$.*

Proof Let K be the minimal splitting field of $f(x)$ and G the Galois group of K over $\mathbb{Q}$.

Reading modulo 2, we see that $f(x) \equiv x^2 + x + 1 \pmod 2$ since p is an odd prime. It is easy to see that $f(x)$ is irreducible over $\mathbb{Z}_2$, and thus it is irreducible over $\mathbb{Q}$ too. Therefore $G \cong S_3$, and hence the Galois group G over $\mathbb{Q}(\sqrt{-m})$ is a cubic cyclic group. More precisely, K is a cubic cyclic extension of $\mathbb{Q}(\sqrt{-m})$.

Since the polynomial $f(x)$ is irreducible over $\mathbb{Q}$ and $G \cong S_3$, so that by Proposition 2.2 K over $\mathbb{Q}(\sqrt{-m})$ is unramified. Therefore by Hilbert class field theory, K is a subfield of the Hilbert class field of $\mathbb{Q}(\sqrt{-m})$. This shows that class number of $\mathbb{Q}(\sqrt{-m})$ is divisible by 3. □

Proposition 3.2 *3 divides the class number of the imaginary quadratic field* $\mathbb{Q}(\sqrt{p^{12\ell+2}-4})$.

Proof Let $d = p^{12\ell+2} - 4$. Since p is odd prime, so that $d \equiv 1 \pmod 4$ and thus the fundamental unit of $\mathbb{Q}(\sqrt{d})$ is of the form

$$\epsilon_d = \frac{t + u\sqrt{d}}{2}.$$

One checks that $\text{Norm}(\epsilon_d) = 1$ gives the Pell-type equation

$$t^2 - u^2 d = 4.$$

We see $t = p^{6\ell+1}$ and $u = 1$ is the smallest solution of this equations, and hence $\varepsilon_d = \dfrac{p^{6\ell+1} + \sqrt{d}}{2}$ is the fundamental unit in $\mathbb{Q}(\sqrt{d})$.

Since $u = 1$, so that Propositions 2.1 and 3.1 together conclude that $h(d) \equiv 0 \pmod 3$. This completes the proof. □

We now give the proof of Theorem 1.1. Propositions 3.1 and 3.2 together complete the proof of the first part. It remains to show that the infinitude of the pair

$$\left(\mathbb{Q}(\sqrt{p^{12\ell+2}-4}), \mathbb{Q}\left(\sqrt{\frac{4-p^{12\ell+2}}{3}}\right)\right).$$

It is easy to see that there are infinitely many prime numbers of the form $p \equiv \pm 4 \pmod 9$ satisfying $m = \dfrac{p^{12\ell+2}-4}{3} \neq \square$. We assume that $\mathfrak{S}$ is the set of all such primes that give the same field more than once. Thus if $p_1 \in \mathfrak{S}$, then one can write

$$\frac{p_0^{12\ell+2} - 4}{3} = Db^2,$$

where b is an integer and D is a square-free positive integer. Therefore if

$$\mathbb{Q}\left(\sqrt{-\left(\frac{p^{12\ell+2}-4}{3}\right)}\right) = \mathbb{Q}\left(\sqrt{-\left(\frac{p_0^{12\ell+2}-4}{3}\right)}\right),$$

then there are integers x and y with $\gcd(x, y) = 1$ such that

$$\left(\frac{p^{12\ell+2}-4}{3}\right)x^2 = Db^2y^2.$$

This shows that $\left(p, \frac{by}{x}\right)$ is an integral on the curve

$$X^{12\ell+2} = 3DY^2 + 4.$$

For any positive integer D, this is an irreducible algebraic curve (see [16]) of genus bigger than 0. From Siegel's theorem (see [7, 17]), it follows that there are only finitely many integral points (X, Y) on this curve. Therefore S is finite and hence there are infinitely many imaginary quadratic fields of the form $\mathbb{Q}(\sqrt{-m})$, where $3m = p^{12\ell+2} - 4$. Again corresponding to each $\mathbb{Q}(\sqrt{-m})$, one gets a real quadratic field $\mathbb{Q}(\sqrt{p^{12\ell+2} - 4})$. Thus, we complete the proof of Theorem 1.1.

Acknowledgements The authors would like to thank Azizul Hoque for suggesting the problem and also for a careful reading of this manuscript. The authors are also indebted to the anonymous for the thorough perusal of this paper.

References

1. N. Ankeny, E. Artin, S. Chowla, The class number of real quadratic fields. Ann. Math. **56**, 479–493 (1952)
2. K. Chakraborty, A. Hoque, Y. Kishi, P.P. Pandey, Divisibility of the class numbers of imaginary quadratic fields. J. Number Theory **185**, 339–348 (2018)
3. K. Chakraborty, A. Hoque, Class groups of imaginary quadratic fields of 3-rank at least 2. Ann. Univ. Sci. Budapest. Sect. Comput. **47**, 179–183 (2018)
4. K. Chakraborty, A. Hoque, R. Sharma, Divisibility of class numbers of quadratic fields: Qualitative aspects, in *Advances in Mathematical inequalities and Application* (Trends Math, Birkhäuser/Springer, Singapore, 2018), pp. 247–264
5. K. Chakraborty, A. Hoque, Divisibility of class numbers of certain families of quadratic fields. J. Ramanujan Math. Soc. **34**(3), 281–289 (2019)
6. K. Chakraborty, A. Hoque, Exponents of class groups of certain imaginary quadratic fields. Czechoslovak Math. J. (2019) (to appear)
7. J.-H. Evertse, J.H. Silverman, Uniform bounds for the number of solutions to $Y^n = f(X)$. Math. Proc. Cambridge Philos. Soc. **100**, 237–248 (1986)
8. A. Hoque, H.K. Saikia On generalized Mersenne primes and class-numbers of equivalent quadratic fields and cyclotomic fields. SeMA J. **67**, 71–75 (2015)
9. A. Hoque, H.K. Saikia, A note on quadratic fields whose class numbers are divisible by 3. SeMA J. **73**, 1–5 (2016)
10. H. Ichimura, Note on the class numbers of certain real quadratic fields. Abh. Math. Sem. Univ. Hamburg **73**, 281–288 (2003)
11. A. Ito, Remarks on the divisibility of the class numbers of imaginary quadratic fields $\mathbb{Q}(\sqrt{2^{2k} - q^n})$. Glasgow Math. J. **53**, 379–389 (2011)
12. Y. Kishi, K. Miyake, parameterization of the quadratic fields whose class numbers are divisible by three. J. Number Theory **80**, 209–217 (2000)
13. Y. Kishi, Note on the divisibility of the class number of certain imaginary quadratic fields. Glasgow Math. J. **51**, 187–191 (2009)

14. S.R. Louboutin, On the divisibility of the class number of imaginary quadratic number fields. Proc. Amer. Math. Soc. **137**, 4025–4028 (2009)
15. T. Nagell, Über die Klassenzahl imaginär quadratischer, Zählkörper. Abh. Math. Sem. Univ. Hamburg. **1**, 140–150 (1922)
16. W.M. Schmidt, equations over finite fields, an elementary approach, in *Lecture Notes in Math* vol. 536 (Springer, Berlin, New York, 1976)
17. C.L. Siegel, Uber einige Anwendungen Diophantischer Approximationen, Abh. Preuss. Akad. Wiss. Phys. Math. Kl. 1, 1–70; Ges. Abh. **1**, 209–266 (1929)
18. K. Soundararajan, Divisibility of class numbers of imaginary quadratic fields. J. Lond. Math. Soc. **61**, 681–690 (2000)
19. K. Uchida, Unramified extension of quadratic number fields, II. Tohoku Math. J. **22**, 220–224 (1970)
20. Y. Yamamoto, On unramified Galois extensions of quadratic number fields. Osaka J. Math. **7**, 57–76 (1970)
21. P.J. Weinberger, Real Quadratic fields with Class number divisible by n. J. Number theory **5**, 237–241 (1973)

On Lebesgue–Ramanujan–Nagell Type Equations

Richa Sharma

2010 Mathematics Subject Classification Primary: 11D61 · Secondary: 11D41

1 Introduction

Mixed polynomial–exponential type equations are of classical and current interest. These equations have been studied extensively by several mathematicians over the years, and thus there are many interesting and fundamental results. These equations appear in most of the topics of numbers theory; in particular. in the study of class numbers/class groups of number fields. In this article, we would like to survey some results concerning the solvability/insolvability of

$$x^2 + D^m = \lambda y^n, \tag{1.1}$$

where D and λ are fixed positive integers, in positive integers x, y, m and n when $\lambda = 1, 2, 4$. We will also provide a sketch of the proof of some important results. This is by no means a complete survey along this topic, and thus it may miss many references and interesting results.

When $y = A$ is fixed in (1.1), the resulting equation is known as a Ramanujan–Nagell type equation, viz.,

$$x^2 + D = \lambda A^n. \tag{1.2}$$

We now recall the following result of Siegel [77] which gives the finiteness of the number of solutions in x and n of (1.2).

R. Sharma (✉)
Malaviya National Institute of Technology Jaipur, Jaipur 302017, India
e-mail: richasharma582@gmail.com

K. Chakraborty et al. (eds.), *Class Groups of Number Fields and Related Topics*,
https://doi.org/10.1007/978-981-15-1514-9_14

Theorem 1.1 *Let $f(x) \in \mathbb{Z}[X]$, and $P(x)$ denotes the greatest prime factor of a positive integer x. If $f(x)$ has at least two distinct roots, then $P(f(x)) \to \infty$ as $|x| \to \infty$.*

The famous Ramanujan–Nagell equation,

$$x^2 + 7 = 2^n \tag{1.3}$$

is a particular case of (1.2). It was Ramanujan [73] who conjectured that all the solutions (x, n) of (1.3) are given by

$$(x, n) \in \{(1, 3), (3, 4), (5, 5), (11, 7), (181, 15)\}.$$

Ljunggren stated the same problem in 1943. It was Nagell who settled it in 1948. Some alternative proofs have been offered by some other authors. For instance, Chowla et al. [30] gave a proof using Skolem's p-adic method.

We now consider the case when $\lambda = 1$ and $A = p$ an odd prime, that is the following:

$$x^2 + D = p^n, \qquad p \nmid D. \tag{1.4}$$

In 1960, Apéry [1] proved that (1.4) has at most two solutions. Further Beukers [11, 12] proved that (1.4) has at most one solution with some exceptions. These exceptions are

$$(p, D) \in \{(3, 2), (4t^2 + 1, 3t^2 + 1)\}$$

for a positive integer t. For these exception cases, there are exactly two solutions. Using hypergeometric methods, Beukars proved these results.

2 The Equation $x^2 + D^m = y^n$

In this section, we discuss some important results concerning the solutions in positive integers of (1.1) when $\lambda = 1$. More precisely, we consider the following:

$$x^2 + D^m = y^n. \tag{2.1}$$

Many special cases of (2.1) for $m = 1$, that is the following equation

$$x^2 + D = y^n \tag{2.2}$$

have been considered by many authors, but most results for general n are of recent origin. Basically (2.1) is known as the generalized Ramanujan–Nagell equation. This is a kind of exponent-type Diophantine equation and has been studied extensively in recent times. When $n = 3$, it is an elliptic curve. Mordell studied this type of equation more carefully and illustrated most of the important results in his book [59]. For $n > 3$, it is a hyperelliptic curve which seems to be more difficult to study. But there is also a vast amount of literature.

Fermat showed that when $D = 2$ and $n = 3$, the only solution is given by $(x, y) = (5, 3)$; a proof was published by Euler in [35]. The first result for general n was due to Lebesgue [48] who proved that when $D = 1$, there is no solution. It was Nagell [65] who proved that there is no solution for $D = 3$ and 5, but did not complete the proof for $D = 2$. He also generalized Fermat's result in [66] and showed that for $D = 2$ the equation has no solution other than $x = 5$. A result rediscovered by himself in [67], who also showed in [68] that when $D = 4$, the only solution is due to $x = 2$ and $x = 11$. Chao [29] proved that $x = 3$ gives the only solution for $D = -1$, a result which had been sought for many years as a case of the Catalan conjecture; so-called Mihăilescu's theorem. The case when n is even can easily be treated, since then D is to be expressed as the difference of two integer squares. On the other hand, for the case n odd, there is no loss of generality in considering the only odd primes p, one has to consider the positive values of D. For $D = 3$, Cohn (2.2) proved that it has no solution in positive integers x,y and $n \geq 3$. There are only a few results for general $n = p$ in (2.2). Blass in [16], and Blass and Steiner jointly in [17] proved (2.2) for $n = 5$ and 7. Cohn [32] gave a brief summary of (2.2). He developed a method by which he found the solutions for 77 values of D up to 100. His methods are inventive but elementary, in the sense that they do not rest on deep tools from Diophantine approximation. The smallest value of D not treated by Cohn is $D = 7$. The difficulty comes from the fact that $8 = (1 + \sqrt{-7})(1 - \sqrt{-7})$ in the imaginary quadratic field $\mathbf{Q}(\sqrt{-7})$. Mignotte and de Weger [58] solved (2.2) for $D = 74$ and obtained $(x, y, n) = (13, 3, 5), (985, 99, 3)$. They also proved that (2.2) has no solution for $D = 86$. Indeed, Cohn solved these two equations of type (2.2) except for $p = 5$, in which difficulties occur since the class numbers of the corresponding imaginary quadratic fields are divisible by 5. Bennett and Skinner [8] used theory of Galois representations and modular forms to solve completely the case when $D = 55$. They obtained $(x, y, n) = (3, 2, 6), (3, 4, 3)$,$(419, 56, 3)$. They also obtained $(x, y, n) = (11, 6, 3), (529, 6, 7)$ when $D = 95$. The 19 remaining values, namely $D = 7, 15, 18, 23, 25, 28, 31, 39, 45, 47, 60, 63, 71, 72, 79, 87, 92, 99, 100$ are clearly beyond the scope of Cohn's method. Most of these cases were considered by Bugeaud, Mignotte, and Siksek and solved in [20]. Note that Cohn [32] proved that there is no solution at all for 46 values of D, namely, 1, 3, 5, 6, 8, 9, 10, 14, 21, 22, 24, 27, 29, 30, 33, 34, 36, 37, 38, 41, 42, 43, 46, 50, 51, 52, 57, 58, 59, 62, 66, 68, 69, 70, 73, 75, 78, 82, 84, 85, 88, 90, 91, 93, 94, and 98. The 31 values of D for which (2.2) has solutions are given in Table 1 with the values of x as well.

Bugeaud, Mignotte, and Siksek also proved that the only solution in positive integers x, y and $n \geq 3$ of (2.2) are given by

$$(x, y, n) \in \{(1, 2, 3), (3, 2, 4), (5, 2, 5), (11, 2, 7), (181, 2, 15)\}.$$

Table 1 Values of D for which (2.2) has solutions

C	x	C	x	C	x	C	x
2	5	20	14	53	26,156	77	2
4	2,11	26	1,207	54	17	80	1
11	4,58	32	7,88	56	5,76	81	46
12	2	35	36	61	8	83	140
13	70	40	52	64	8	89	6
16	4	44	9	65	4	96	23
17	8	48	4,148	67	110	97	48
19	18,22434	49	24,524	76	7,1015		

However, they showed that (2.2) has no more solutions than that of (1.3) when $D = 7$. Earlier results on $x^2 + 7 = y^n$ were due to Lesage [49] and to Siksek and Cremona [78]. On the other hand, Sury [86] presented an elementary proof of (2.2) for $D = 2$ and $n > 1$ that the only solution is $(x, y, n) = (\pm 5, 3, 3)$. Recently, Hoque and Saikia investigated (2.2) for any positive integer D in [39]. They used this fact to study the class number of associated imaginary quadratic field.

In some recent papers, some more complicated cases where D is a product of more than one prime powers have been considered. Let

$$S = \{p_1, p_2, \dots, p_s\}$$

denote a set of s-distinct primes and $\mathfrak{S}$ the set of nonzero integers composed only of the primes from S. Let P be the maximum element from S and $\mathcal{P}$ the product of the primes of S. In recent years, (2.2) has been considered also in the more general case, when D is no longer fixed but $D \in \mathfrak{S}$ with $D > 0$. It follows from [75, Theorem 2] that n in (2.2) can be bounded from above by a computable constant depending only on P and s. In [38], an effective upper bound was derived for n which depends only on $\mathcal{P}$.

Luca solved (2.2) completely in [55] when $D = 2^a 3^b$ with $n \geq 3$ and $\gcd(x, y) = 1$. The solutions are as follows:

$$(x, y) = \begin{cases} (7, 3), (23, 5), (7, 5), (47, 7), (287, 17) \text{ if } n = 4, \\ (5, 3), (11, 5), (10, 7), (17, 7), (46, 13), (35, 13), (595, 73), \\ (995, 97), (2681, 193), (39151, 1153) \text{ if } n = 3. \end{cases}$$

Muriefah, Luca, and Togbé determined all solutions of (2.2) for the case $D = 5^a 13^b$ with $\gcd(x, y) = 1$. In this case, all the solutions are given by

$$(x, y, a, b) = \begin{cases} (70, 17, 0, 1), (142, 29, 2, 2) \text{ if } n = 3, \\ (4, 3, 1, 1) \text{ if } n = 4. \end{cases}$$

Cangul, Demirci, Luca, Pinter, and Soydan [21] obtained all solutions of (2.2) for $D = 2^a 11^b$ with $\gcd(x, y) = 1$. The case $D = 2^a 3^b 11^c$ was also considered by Cangul, Demirci, Inam, Luca, and Soydan in [22]. In this case, the solutions of (2.2) are given with the condition $\gcd(x, y) = 1$. The complete solution of (2.2) when $D = 5^a 11^b$ with $\gcd(x, y) = 1$ has been obtained by Cangul, Demirci, Soydan, and Tzanakis [23] with some exceptions. Pink and Rabai [72] determined all solutions of (2.2) when $D = 5^a 17^b$ with $\gcd(x, y) = 1$. Godinho, Marques, and Togbé [36] solved completely (2.2) when $D = 2^a 5^b 17^c$ under the assumption $\gcd(x, y) = 1$. Soydan [81] found all the solutions of (2.2) with $D = 7^a 11^b$ for the nonnegative integers $a, b, x, y, n \geq 3$, where x and y are coprime, except when a, x are odd and b is even. Cenberci and Peker [24] investigated the case $D = 19^m$, while Soydan, Ulas, and Zhu [82] solved completely (2.2) when $D = 2^a 19^b$ under the assumption of $\gcd(x, y) = 1$. Soydan [83] proved that if $D = \pm 5^\alpha p^n$, then (2.2) has no solutions in positive integers α,x,y with $\gcd(x, y) = 1$, x is odd, $n \geq 7$, and $p \notin \{2, 5\}$. Pink [70] obtained all the non-exceptional solutions of (2.2) for the case $D = 2^\alpha 3^\beta 5^\gamma 7^\delta$. Godinho, Marques, and Togbe [37] found all positive integer solutions of (2.2) with $D = 2^a 3^b 17^c$ and $D = 2^a 13^b 17^c$, $\gcd(x, y) = 1$. In [74], Saradha and Srinivasan discussed (2.2) for $p_l', \cdots, p_r^r = D_s D^{t^2}$, where D_s is the square-free part of D and $\alpha_1, \alpha_2, \cdots, \alpha_r$ are positive integer unknowns. They obtained several interesting results concerning some of the values of $D_s \leq 10000$. Further, for the case $D = p^l$, $p \in \{11, 19, 43, 67\}$, it was proved in the same paper that (2.2) may have a solution only if $D = 3^\beta 5^\gamma$. Le and Zhu [47] solved completely (2.2) when $D = p^l$ with $p \in (11, 19, 43, 67)$, where the class number $h(-d)$. Here $h(-d) = 1$ denotes the class number of the imaginary quadratic field $\mathbb{Q}(\sqrt{-p})$. In [89], Zhu discussed (2.2) when $D = p^a$, where p a prime and $n = 3$. For $D = p^a$, we refer to Bugeaud [18]. Berczes and Pink [10] extending the above result of Saradha and Srinivasan and Le and Zhu, solved completely (2.2) for $D = d^{2l+1}$ in the case $h(-d) \in \{2, 3\}$, where $d > 0$ is a square-free integer.

Recently, several authors became interested in the case when only the prime factors of D are specified. For example, the case when $D = p^k$ with a fixed prime number p. Arif and Muriefah [2] solved $x^2 + 2^k = y^n$ under certain assumptions. Cohn [31] considered the Diophantine equation $x^2 + 2^k = y^n$, where $n \geq 3$ and k was supposed odd, and demonstrated that there were exactly three families of solutions. The same problem with even k appears to be of rather greater difficulty and was considered by Arif and Abu Muriefah [2]. They made the following conjecture:
If $k = 2m$, the diophantine equation $x^2 + 2^k = y^n$ has precisely two families of solutions, given by $x = 2^m$ for all m and by $n = 3$, $x = 11.2^{3M}$ if $m = 3M + 1$.

Arif and Abu Muriefah gave a partial answer to this conjecture in [2] and also did Cohn in [34]. However, they finally proved the conjecture in [4]. Le [46] and Siksek [79] presented an alternative proof of the same. Le [46] verified a conjecture of Cohn [31] by determining all the solutions of the Diophantine equation $x^2 + 2^k = y^n$ in positive integers x, y, k, n with $2 \nmid y$ and $n \geq 3$, viz., $(x, y, k, n) = (5, 3, 1, 3), (7, 3, 5, 4), (11, 5, 2, 3)$. Arif and Muriefah [3] proved that the Diophantine equation $x^2 + 3^m = y^n$, $n \geq 3$ has only one solution in positive integers x, y, m and the unique solution is given by $m = 5 + 6M$, $x = 10.3^{3M}$,

$y = 7.3^{2M}$ and $n = 3$, when m is odd. It is also proved that there is no solution when n is even. Luca proved this conjecture in [54]. Luca and Togbé [56] found all the solutions of the equation $x^2 + 7^{2k} = y^n$ where $x \geq 1$, $y \geq 1$, $k \geq 1$ and $n \geq 3$ and solutions are given by

$$(x, y, k) = \begin{cases} (524.7^{3\lambda}, 65.7^{2\lambda}, 1 + 3\lambda) \text{ if } n = 3, \\ (24.7^{2\lambda}, 5.7^{\lambda}, 1 + 2\lambda) \text{ if } n = 4 \text{ where } \lambda \geq 0 \text{ is any integer.} \end{cases}$$

In 2008, Liqun [50] found all the solutions of the equation $X^2 + 3^m = Y^n$ with the help of a deep result due to Bilu, Hanrot, and Voutier. In [51], he proved that the Diophantine equation $x^2 + 5^m = y^n, n > 2, m > 0$ has no positive integer solution when $2 \nmid m$, nor when $2 \mid m$ under the additional condition $\gcd(x, y) = 1$, with the help of Bilu, Hanrot, and Voutier's method. Muriefah [63] established that the equation $x^2 + 5^{2k} = y^n$, where $k > 0$ and $n > 3$, may have a solution in integers (x, y, k, n) only when $5 \mid x$ and $p \nmid k$, where p any odd prime dividing n, by using a recent method of Bilu, Hanrot, and Voutier [14]. Finally, Muriefah and Arif [61] proved that the equation $x^2 + 5^{2k+1} = y^n, n \geq 3$ has no solution in positive integer x,y for all $k \geq 0$. Several results have been also obtained by Muriefah and Arif in [62] for $D = q^{2k}$, where q is an odd prime. The equation (2.1) is naturally well connected with the investigation of the class number of the imaginary quadratic number field $\mathbb{Q}(\sqrt{-D})$. The solvability of some special cases of (2.1) has been used in [25–27, 40] to investigate the class numbers of associated imaginary quadratic number fields. Arif and Muriefah [5] determined all the solutions of the equation $x^2 + q^{2k+1} = y^n$, when q is an odd prime, $q \not\equiv 7 \pmod 8$, n is an odd integer ≥ 5, n is not a multiple of 3 and $\gcd(n, h) = 1$, where h is the class number of the field $\mathbb{Q}(\sqrt{(-q)})$ has exactly two families of solutions given by Table 2

Zhu discussed $x^2 + q^m = y^3$ completely in [89]. Soydan, Demirci, and Cangul [80] found all the solutions of the equation $x^2 + 11^m = y^n$, in positive integers x,y with odd $m > 1$ and $n \geq 3$. Berczes and Pink [9] solved the equation $x^2 + p^{2k} = y^n$, where $2 \leq p \leq 100$ is a prime and integer unknowns x, y, n, k satisfying $x > 0$, $y > 1, n \geq 3$ prime, $k \lesssim 0$, and $\gcd(x, y = 1$. They also proved that there is no solution of the equation $x^2 + p^{2k} = y^p$ in integer unknowns (x, y, p, k) with $x \geq 1$, $y > 1$, $p \geq 5$ prime, $k \geq 0$, and $\gcd(x, y) = 1$. Peker and Cenberi [71] proved that the equation $x^2 + 19^m = y^n$ has solution for not only $2 \mid m$ but also $2 \nmid m$. One can remark that many cases of (2.1) are yet to be considered.

Table 2 Solutions of $x^2 + q^{2k+1} = y^n$ given by Arif and Muriefah

x	y	q	k	n
22434×19^{5M}	55×19^{2M}	19	5M	5
2759646×341^{5M}	377×19^{2M}	341	5M	5

3 The Equation $x^2 + D^m = 2y^n$

The Diophantine equation

$$x^2 + D = 2y^n, \tag{3.1}$$

with some restrictions, has received a little attention. Cohn [33] showed that the only solution to (3.1) for $D = 1$ are $x = y = 1$ and $x = 239$, $y = 13$, and $n = 4$. Tengely considered [88] the equation

$$x^2 + q^{2m} = 2y^p, \tag{3.2}$$

where m, p, q, x, and y are integer unknowns with $m > 0$, p and q odd primes and x and y coprime. More precisely, he proved the following:

Theorem 3.1 *There are only finitely many solutions* (x, y, m, q, p) *of* (3.2) *with* $gcd(x, y) = 1$, $x, y \in \mathbb{N}$ *such that* y *is not a sum of two consecutive squares,* $m \in \mathbb{N}$, *and* $p > 3$, q *are odd primes.*

In order to prove this theorem, he proved the following proposition which provides bounds for p.

Proposition 3.1 *If* $x^2 + q^{2m} = 2y^p$ *admits a relatively prime solution* $(x, y) \in \mathbb{N}^2$, *then*

$$(x, y) = \begin{cases} p \le 3803 \ \ if \ \ u + \delta_4 v = \pm q^m, q^m \ge 503, \\ p \le 3089 \ \ if \ \ p = q, \\ p \le 1309 \ \ if \ \ u + \delta_4 v = \pm q^m, m \ge 40, \\ p \le 1093 \ \ if \ \ u + \delta_4 v = \pm q^m, m \ge 100, \\ p \le 1009 \ \ if \ \ u + \delta_4 v = \pm q^m, m \ge 250. \end{cases}$$

He also proved

Theorem 3.2 *The only solution* (m, p, q, x) *in positive integers* m, p, q, *and* x *with* p, q *odd primes, of the equation*

$$x^2 + q^{2m} = 2.17^p$$

is $(1, 3, 5, 99)$.

He also solved (3.2) when $q = 3$. More precisely, he proved

Theorem 3.3 *If the equation*

$$x^2 + 3^{2m} = 2y^p$$

with $m > 0$ *and* p *prime admit a coprime integer solution* (x, y), *then*

$$(x, y, m, p) \in \{(13, 5, 2, 3), (79, 5, 1, 5), (545, 53, 3, 3)\}.$$

Proposition 3.2 *There exists no coprime integer solution* (x, y) *of*

$$x^2 + 3^{2m} = 2y^p$$

with $m > 0$ *and* $p < 1000$, $p \equiv 5 \pmod{24}$ *or*

$$p \in \{131, 251, 491, 971\}$$

prime.

The Thue equations related to the remaining primes $p < 1000$ were solved by G. Hanrot.

Proposition 3.3 (G. Hanrot) *There exists no coprime integer solution* (x, y) *of*

$$x^2 + 3^{2m} = 2y^p$$

with $m > 0$ *and*

$$p \in \{59, 83, 107, 179, 227, 347, 419, 443, 467, 563, 587, 659, 683, 827, 947\}.$$

Muriefah, Luca, Siksek, and Tengely [64] investigated (3.1) both in the case where D is a fixed integer as well as in the case where D is the product of powers of fixed primes. They showed that in some cases, this equation can be solved by appealing to the theorem of Bilu, Hanrot, and Voutier [14] on primitive divisors of Lehmer sequences. More precisely, they prove the following results.

Theorem 3.4 *Let* $D > 0$ *be an integer satisfying* $D \equiv 1 \pmod 4$ *and write* $D = cd^2$, *where* c *is square-free. Suppose that* (x, y) *is a solution of* (3.1) *with* $x, y \in \mathbb{Z}^+$, $gcd(x, y) = 1$ *and* $p \geq 5$ *a prime. Then one of the following is true:*

(i) $x = y = D = 1$,
(ii) p *divides the class number of the quadratic field* $\mathbb{Q}(\sqrt{-d})$,
(iii) $p = 5$ *and* (D, x, y) *is one of* $(9, 79, 5)$, $(125, 19, 3)$, $(125, 183, 7)$, *and* $(2125, 21417, 47)$,
(iv) $p \mid \left(q - \left(\frac{-d}{q}\right)\right)$, *where* q *is some odd prime such that* $q \mid d$ *and* $q \nmid c$. *Here* $\left(\frac{-d}{q}\right)$ *denotes the Legendre symbol of the integer* $-d$ *with respect to the prime* q.

Theorem 3.5 *The only solutions of* (3.1) *with* x, y *coprime integers,* $n \geq 3$, *and* $D \equiv 1 \pmod 4$, $1 \leq D \leq 100$ *are* $1^2 + 1 = 2\Delta 1^n$, $79^2 + 9 = 2\Delta 5^5$, $5^2 + 29 = 2\Delta 3^3$, $117^2 + 29 = 2\Delta 19^3$, $993^2 + 29 = 2\Delta 79^3$, $11^2 + 41 = 2\Delta 3^4$, $69^2 + 41 = 2\Delta 7^4$, $171^2 + 41 = 2\Delta 11^4$, $1^2 + 53 = 2\Delta 3^3$, $25^2 + 61 = 2\Delta 7^3$, $51^2 + 61 = 2\Delta 11^3$, $37^2 + 89 = 2\Delta 9^3$.

As a consequence of Theorem 3.4, one gets the following:

Corollary 3.1 *Let $q_1, \dots q_k$ be distinct primes satisfying $q_i \equiv 1 \pmod 4$. Suppose that $(x, y, p, a_1, \dots a_k)$ is a solution to the equation*

$$x^2 + q_1^{a_1} \dots q_k^{a_k} = 2y^p,$$

satisfying $x, y \in \mathbb{Z}^+$, $gcd(x, y) = 1$, $a_i \geq 0$, $p \geq 5$ prime. Then one of the following is true:

(i) $x = y = 1$ *and all the* $a_i = 0$,
(ii) p *divides the class number of the quadratic field* $\mathbb{Q}(\sqrt{-d})$ *for some square-free* d *dividing* $q_1 q_2 \dots q_k$,
(iii) $p = 5$ *and* $(\prod q_i^{a_i}, x, y) = (125, 19, 3), (125, 183, 7), (2125, 21417, 47)$,
(iv) $p \mid (q_i^2 - 1)$ *for some* i.

Muriefah, Luca, Siksek, and Tengely solved completely the following equations under the restrictions $\gcd(x, y) = 1$ and $n \geq 3$:

$$x^2 + 17^{a_1} = 2y^n,$$

$$x^2 + 5^{a_1} 13^{a_2} = 2y^n, \text{ and}$$

$$x^2 + 3^{\alpha_1} 11^{\alpha_2} = 2y^n.$$

They also proved when $d \not\equiv 1 \pmod 4$, one may try to use the modular approach [8] to solve the equation.

in [87], Tengely gave a method to resolve the equation

$$x^2 + a^2 = 2y^n$$

in integers $n > 2$, x, y for any fixed a. He determined all solutions for odd a with $3 \leq a \leq 501$.

Zhu et al. [90] described all the solutions of the equation

$$x^2 + p^{2m} = 2y^n$$

with $\gcd(x, y) = 1$, $n > 2$, where p be an odd prime. They also proved that this equation has no solution (x, y, m, n) when $n > 3$ is an odd prime and y is not the sum of two consecutive squares. This extends the work of Tengely in [88].

However, in 1942, Ljunggren gave a very complicated proof of the fact that the only positive integer solutions of the equation

$$X^2 + \ell = 2Y^4$$

are $(X, Y) = (1, 1)$ and $(239, 13)$. Later, Steiner and Tzanakis [84] gave a simpler solution of Ljunggren's problem. This is accomplished by reducing the problem to a Thue equation and then solving it by using a deep result of Mignotte and Waldschmidt on linear forms in logarithms and continued fractions.

4 On the Equation $x^2 + D^m = 4y^n$

There are many results in the literature concerning various extensions and generalizations of the following equation:

$$x^2 + D^m = 4y^n. \tag{4.1}$$

In 1972, Ljunggren [53] proved some results concerning the solvability in positive integers x, y, and q of (4.1) for a fixed positive integer $D \equiv 3 \pmod 4$ without any square factor greater than 1 when $n = q$ is an odd prime and $m = 1$.

Luca, Tengely, and Togbé [57] determined that the only integer solutions (D, n, x, y) of (4.1) with $x, y \geq 1$, $\gcd(x, y) = 1$, $n \geq 3$, $D \equiv 3 \pmod 4$, $1 \leq D \leq 100$ are given in Table 3.

They also determined the integer solution of (4.1) when $D = 5^a 11^b, 7^a 13^b$ with $x, y \geq 1$, $gcd(x, y) = 1$, $n \geq 3$ and $a, b \geq 0$. In [43], Le proved that if $n > 8.5.10^6$, then the equations

$$d_1 x^2 + 2^{2m} d^2 = y^n \text{ with } 2 \nmid y$$

and

$$d_1 x^2 + d_2 = 4y^n$$

have no positive integer solutions (x, y) with $\gcd(x, y) = 1$. Le [45] also showed in his note that the generalized Ramanujan–Nagell equation

Table 3 Solutions of (4.1) when $m = 1$ and $D \equiv 3 \pmod 4$, $1 \leq D \leq 100$

(3,n,1,1)	(3,3,37,7)	(7,3,5,2)	(7,5,11,2)
(7,13,181,2)	(11,5,31,3)	(15,4,7,2)	(19,7,559,5)
(23,3,3,2)	(23,3,29,6)	(23,3,45,8)	(23,3,83,12)
(23,3,7251,236)	(23,9,45,2)	(31,3,1,2)	(31,3,15,4)
(31,3,63,10)	(31,3,3313,140)	(31,6,15,2)	(35,4,17,3)
(39,4,5,2)	(47,5,9,2)	(55,4,3,2)	(59,3,7,3)
(59,3,21,5)	(59,3,525,41)	(59,3,28735,591)	(63,4,1,2)
(63,4,31,4)	(63,8,31,2)	(71,3,235,24)	(71,7,21,2)
(79,3,265,26)	(79,5,7,2)	(83,3,5,3)	(83,3,3785,153)
(87,3,13,4)	(87,3,1651,88)	(87,6,13,2)	(99,4,49,5)

$$D_1x^2 + D_2 = 4p^n$$

has at most two positive integer solutions (x, n) except $(D_1, D_2, p) = (1, 7, 2)$, $(3, 5, 2)$, $(1, 11, 3)$, and $(1, 19, 5)$, where D_1, D_2 are positive integers with $2 \nmid D_1D_2$ and $\gcd(D_1, D_2) = 1$ and p be a prime with $p \nmid D_1D_2$. Le and Hunan [44] proved that the equation

$$D_1x^2 + D_2^m = 4y^n$$

with $D_1,\ D_2, x, y, m, n \in \mathbb{N}$, $\gcd(D_1x, D_2y) = 1$, $2 \nmid m$, n an odd prime,$n \nmid h$ $(-D_1D_2)$, has only a finite number of solutions (D_1, D_2, x, y, m, n) with $n > 5$. Moreover, the solutions satisfy $4y^n < \exp \exp 470$, where $D_1, D_2 \in \mathbb{N}$ and $h(-D_1D_2)$ denote the class number of the imaginary quadratic field $\mathbb{Q}(\sqrt{-D_1D_2})$. Many special cases of the diophantine equation $ax^2 + b^m = 4y^n$ where $(ax, by) = 1$, a, b are square-free integers, $y > 1$, m is odd, a, b, x, y, m, solved a family and n be positive integers and n is an odd prime, have been considered in the last few years (see [15, 19, 42, 85]). Le [43, 45] studied this equation in full generality and proved that it has only a finite number of solutions (a, b, x, y, m, n) with $n > 5$. Hua and Voutier [41] used the method of Thue and Siegel, based on explicit Pade approximations to algebraic functions, to completely solve a family of quartic Thue equations. Using this result, they considered the Diophantine equation

$$X^2 + 1 = dY^4$$

and proved that this equation has at most one solution in positive integers when $d \geq 3$. Arif and Al-Ali [6] showed that the Diophantine equation

$$ax^2 + b^{2k+1} = 4y^n, \tag{4.2}$$

where a, b, x, y, k, n are positive integers such that $(ax, b) = 1$, a, b are square free integers, $k \geq 0$, n is an odd prime, $(n, h) = 1$ where h is the class number of the field $\mathbb{Q}\left(\sqrt{-ab}\right)$ and $y > 1$, has no solutions in (a, b, x, y, k, n) when $n > 13$ and has exactly six solutions for $7 \leq n \leq 13$, given by $(a, b, k, n, y) = (1, 7, 0, 13, 2), (1, 7, 1, 7, 2), (1, 19, 0, 7, 5),\ \ (3, 5, 0, 7, 2), (5, 7, 1, 7, 3), (13, 3, 0, 7, 4)$. Further if $a = 1$, $n = 5$, then 4.2 has exactly 2 solutions given by $k = 0$ and $(b, y) = (7, 2), (11, 3)$.

Bugeaud [19] applied a new, deep result of Bilu, Hanrot, and Voutier to solve completely some exponential Diophantine equations of the type $D_1x^2 + D_2 = \lambda^2 y^p$, where D_1,D_2 are given coprime positive integers, $\lambda \in \{1, 2\}$ and x, y, and $p \geq 3$ are unknown.

Bhatter, Hoque, and Sharma [13] completely solved (4.1) for any nonnegative integer x,y,m, and n when $D = -19$. More precisely, they proved.

Theorem 4.1 *Let $k \geq 0$ be an integer. The Diophantine equation*

$$x^2 + 19^{2k+1} = 4y^n \tag{4.3}$$

has no solutions in positive integers x, y, n except

$$(x, y, n) \in \left\{ \left(19^t \times \frac{19^{2(k-t)+1} - 1}{2}, 19^t \times \frac{19^{2(k-t)+1} + 1}{4}, 2 \right), (559 \times 19^{7m}, 5 \times 19^{2m}, 7) \right\}$$

with $t, m \in \mathbb{Z}_{\geq 0}$ satisfying $k = 7m$ in case of the second solution, and $n \neq 1$. For $n = 1$, it has infinitely many solutions in positive integers x, y, n.

Recently, Chakraborty et al. [28] completely solved (4.1) when $-D$ is one of $7, 11, 19, 43, 67, 163$.

Acknowledgements The author would like to thank the referee whose suggestions and comments helped to improve the manuscript. The author would like to thank Harish-Chandra Research Institute and Malaviya National Institute of Technology, Jaipur for providing sufficient facility to prepare this manuscript.

References

1. R. Apéry, Sur ue équation Diophantienne, C. R. Acad. Sci. Paris Sér. A **251**(1263–1264), 1451–1452 (1960)
2. S. A. Arif and F. S. Abu Muriefah, On the Diophantine equation $x^2 + 2^k = y^n$, Int. J. Math. Sci. **20**, 299–304 (1997)
3. S.A. Arif, F.S. Abu Muriefah, The Diophantine equation $x^2 + 3^m = y^n$, Int. J. Math. Math. Sci. **21**, 610–620 (1998)
4. S.A. Arif, F.S. Abu Muriefah, On the Diophantine equation $x^2 + 2^k = y^n$. II, Arab. J. Math. Sci. **7**, 67–71 (2001)
5. S. A. Arif, F.S. Abu Muriefah, On the Diophantine equation $x^2 + q^{2k+1} = y^n$, J. Number Theory **95**, 95–100 (2002)
6. S.A. Arif, A.S. Al-Ali, On the Diophantine equation $ax^2 + b^m = 4y^n$. Acta Arithmetica **103**(4), 343–346 (2002)
7. S.A. Arif, F.S. Abu Muriefah, On the Diophantine equation $x^2 + 5^{2k} = y^n$, Demonstratio Math. **319**(2), 285–289 (2006)
8. M.A. Bennett, C.M. Skinner, Ternary Diophantine equations via Galois representations and modular forms. Canad. J. Math. **56**, 23–54 (2004)
9. A. Berczes, I. Pink, On the Diophantine equation $x^2 + p^{2k} = y^n$. Arch. Math. **91**, 505–517 (2008)
10. A. Berczes, I. Pink, On the Diophantine equation $x^2 + d^{2\ell+1} = y^n$. Glasg. Math. J. **54**, 415–428 (2012)
11. F. Beukers, On the generalized Ramanujan-Nagell equations, I, Acta Arith. **38** (1980/1981), 389–410
12. F. Beukers, On the generalized Ramanujan-Nagell equations. II, Acta Arith. **39**, 113–123 (1981)
13. S. Bhatter, A. Hoque, R. Sharma, On the solutions of a Lebesgue-Nagell type equation. Acta Math. Hungar. **158**(3), 17–26 (2019)
14. Y. Bilu, G. Hanrot, P.M. Voutier, Existence of primitive divisors of Lucas and Lehmer numbers. J. Reine Angew. Math. **539**, 75–122 (2001)

15. Y. Bilu, On Le's and Bugeaud's papers about the equation $ax^2 + b^{2m-1} = 4c^p$. Monatsh. Math. **137**(1), 1–3 (2002)
16. J. Blass, A note on diophantine equation $Y^2 + k = X^5$. Math. Comp. **30**, 638–640 (1976)
17. J. Blass, R. Steiner, On the equation $y^2 + k = x^7$. Utilitas Math. **13**, 293–297 (1978)
18. Y. Bugeaud, On the Diophantine equation $x^2 - p^m = \pm y^n$. Acta Arith. **80**, 213–223 (1997)
19. Y. Bugeaud, On some exponential Diophantine equations. Monatsh. Math. **132**, 93–97 (2001)
20. Y. Bugeaud, M. Mignotte, S. Siksek, Classical and modular approaches to exponential Diophantine equations II. The Lebesgue-Nagell equation, Compositio Math. **142**, 31–62 (2006)
21. I.N. Cangul, M. Demirci, F. Luca, A. Pinter, G. Soydan, On the Diophantine equation $x^2 + 2^a 11^b = y^n$. Fibonacci Quart. **48**, 39–46 (2010)
22. I.N. Canqul, M. Demirci, I. Inam, F. Luca, G. Soydan, On the Diophantine equation $x^2 + 2^a 3^b 11^c = y^n$. Math. Slovaca **63**, 647–659 (2013)
23. I.N. Cangul, M. Demirci, G. Soydan, N. Tzanakis, On the Diophantine equation $x^2 + 5^a 11^b = y^n$. Funct. Approx. Comment. Math. **43**, 209–225 (2010)
24. S. Cenberci, B. Peker, On the solutions of the equation $x^2 + 19^m = y^n$. Notes Number Theory Disc. Math. **18**, 34–41 (2012)
25. K. Chakraborty, A. Hoque, Y. Kishi, P.P. Pandey, Divisibility of the class numbers of imaginary quadratic fields. J. Number Theory **185**, 339–348 (2018)
26. K. Chakraborty, A. Hoque, R. Sharma, *Divisibility of class numbers of quadratic fields: Qualitative aspects, Advances in Mathematical inequalities and Application, 247–264* (Trends Math. Birkhäuser/Springer, Singapore, 2018)
27. K. Chakraborty , A. Hoque, *Exponents of class groups of certain imaginary quadratic fields*, Czechoslovak Math. J. (to appear). arXiv:1801.00392
28. K. Chakraborty, A. Hoque, R. Sharma, Complete solutions of certain Lebesgue-Ramanujan-Nagell equations (2019). arxiv:1812.11874
29. K. Chao, On the Diophantine equation $x^2 = y^n + 1, xy \neq 0$. Sci. Sinica (Notes) **14**, 457–460 (1964)
30. S. Chowla, D.J. Lewis, Th Skolem, The Diophantine equation $2^{n+2} - 7 = x^2$ and related problems. Proc. Amer. Math. Soc. **10**, 250–257 (1959)
31. J.H.E. Cohn, The diophantine equation $x^2 + 2^k = y^n$. Arch. Math. (Basel) **59**(4), 341–344 (1992)
32. J.H.E. Cohn, The Diophantine equation $x^2 + C = y^n$. Acta Arith. **55**, 367–381 (1993)
33. J.H.E. Cohn, Perfect Pell powers. Glasgow Math. J. **38**, 19–20 (1996)
34. J.H.E. Cohn, The Diophantine equation $x^2 + 2^k = y^n$. II, Int. J. Math. Math. Sci., **22**, 459–462 (1999)
35. L. Euler, *Algebra*, vol. 2
36. H. Godinho, D. Marques, A. Togbe, On the Diophantine equation $x^2 + 2^{\alpha} 5^{\beta} 17^{\gamma} = y^n$. Comm. Math. **20**, 81–88 (2012)
37. H. Godinho, D. Marques and A. Togbe, On the Diophantine equation $x^2 + C = y^n$ for $C = 2^a 3^b 17^c$ and $C = 2^a 13^b 17^c$, Math. Slovaca **66**(3), 565–574 (2016)
38. K. Gyory, I. Pink, A. Pinter, Power values of polynomials and binomial Thue-Mahler equations. Publ. Math. Debrecen **65**, 341–362 (2004)
39. A. Hoque, H.K. Saikia, On the divisibility of class numbers of quadratic fields and the solvability of diophantine equations. SeMA J. **73**(3), 213–217 (2016)
40. A. Hoque, K. Chakraborty, Divisibility of class numbers of certain families of quadratic fields. J. Ramanujan Math. Soc. **34**(3), 281–289 (2019)
41. C.J. Hua, P. Voutier, Complete solution of the Diophantine equation $X^2 + 1 = dY^4$ and a related family of quartic Thue equations. J. Number Theory **62**, 71–99 (1997)
42. M.H. Le, On the Diophantine equation $x^2 + D = 4p^n$. J. Number Theory **41**, 87–97 (1992)
43. M.H. Le, On the Diophantine equations $d_1 x^2 + 2^{2m} d^2 = y^n$ and $d_1 x^2 + d_2 = 4y^n$. Proc. Amer. Math. Soc. **118**, 67–70 (1993)
44. M.H. Le, On the Diophantine equation $D_1 x^2 + D_2^m = 4y^n$. Monatsh. Math. **120**, 121–125 (1995)

45. M. Le, A note on the number of solutions of the generalized Ramanujan-Nagell equation $D_1x^2 + D_2 = 4p^n$. J. Number Theory **62**, 100–106 (1997)
46. M. Le, On Cohn's conjecture concerning the Diophantine equation $x^2 + 2^m = y^n$. Arch. Math. (Basel) **78**, 26–35 (2002)
47. M. Le, H. Zhu, On some generalized Lebesque-Nagell equations. J. Number Theory **131**, 458–469 (2011)
48. V. A. Lebesgue, Sur l'impossibilité en nombres entiers de lé equation $x^m = y^2 + 1$, Nouvelles Annales des Mathématiques (1), **9**, 178–181 (1850)
49. J.-L. Lesage, Difference entre puissances et carres d'entiers. J. Number Theory **73**, 390–425 (1998)
50. T. Liqun, On the Diophantine equation $x^2 + 3^m = y^n$, Integers: Electronic J. Comb. Number Theory **8**, #A55, 1–7 (2008)
51. T. Liqun, On the Diophantine equation $x^2 + 5^m = y^n$. Ramanujan J. **19**, 325–338 (2009)
52. W. Ljunggren, On the Diophantine equation $x^2 + D = 4y^q$. Monatsh. Math. **75**, 136–143 (1971)
53. W. Ljunggren, New theorems concerning the Diophantine equation $x^2 + D = 4y^q$. Acta Arith. **21**(2), 183–191 (1972)
54. F. Luca, On a Diophantine equation. Bull. Austral. Math. Soc. **61**, 241–246 (2000)
55. F. Luca, On the equation $x^2 + 2^a3^b = y^n$. Int. J. Math. Sci. **29**, 239–244 (2002)
56. F. Luca, A. Togbe, On the Diophantine equation $x^2 + 7^{2k} = y^n$. Fibonacci Quart. **54**, 322–326 (2007)
57. F. Luca, Sz. Tengely, A. Togbe, On the Diophantine equation $x^2 + C = 4y^n$, Ann. Sci. Math. Qübec **33**(2),171–184 (2009)
58. M. Mignotte, B.M.M. de Weger, On the Diophantine equations $x^2 + 74 = y^5$ and $x^2 + 86 = y^5$. Glasgow Math. J. **38**, 77–85 (1996)
59. L.J. Mordell, *Diophantine Equations* (Academic Press, London, 1969)
60. F.S. Abu Muriefah, S.A. Arif, On a Diophantine equation, Bull. Austral. Math. Soc. **57**, 189–198 (1998)
61. F.S. Abu Muriefah, S.A. Arif, The Diophantine equation $x^2 + 5^{2k+1} = y^n$, Indian J. Pure Appl. Math. **30**, 229–231 (1999)
62. F.S. Abu Muriefah, S.A. Arif, The Diophantine equation $x^2 + q^{2k} = y^n$, *Arab. J. Sci. Eng. Sect. A Sci.* **26**, 53–62 (2001)
63. F. S. Abu Muriefah, On the Diophantine equation $x^2 + 5^{2k} = y^n$, Demonstratio Math. **39**(2), 285–289 (2006)
64. F.S. Abu Muriefah, F. Luca, S. Siksek, Sz. Tengely, On the Diophantine equation $x^2 + C = 2y^n$, Int. J. Number Theory **05**(05), 1117–1128 (2009)
65. T. Nagell, Sur l'impossibilité de quelques équations á deux indéterminées. Norsk. Mat. Forensings Skrifter. **13**, 65–82 (1923)
66. T. Nagell, Über einige Arcustangensgleichungen die auf interessante unbestimmte Gleichungen führen, Ark. Mat. Astr. Fys. **29A**(13) (1943)
67. T. Nagell, Verallgemeinerung eines Fermatschen Satzes. Arch. Math. (Basel) **5**, 153–159 (1954)
68. T. Nagell, Contributions to the theory of a category of Diophantine equations of the second degree with two unknowns, Nova Acta Regiae Soc. Sci. Upsaliensis, Ser. 4, **16**(2), 38 pp (1955)
69. B. Persson, On a Diophantine equation in two unknowns. Ark. Mat. **1**, 45–57 (1949)
70. I. Pink, On the Diophantine equation $x^2 + 2^\alpha 3^\beta 5^\gamma 7^\delta = y^n$. Publ. Math. Debrecen **70**(1–2), 149–166 (2007)
71. B. Peker, S. Cenberi, On the solutions of the equation $x^2 + 19^m = y^n$, Notes Number Theory Disc. Math. **18** 2012, no. 2, 34–41
72. I. Pink and Zs. Rabai, On the Diophantine equation $x^2 + 5^k17^l = y^n$, *Comm. Math.* **19**, (2011), 1–9
73. S. Ramanujan, Question 446, *J. Indian Math. Soc.* **5** (1913), 120, Collected papers, Cambridge University Press (1927), 327
74. N. Saradha, A. Srinivasan, Solutions of some generalized Ramanujan-Nagell equations. Indag. Math. (NS) **17**(1), 103–114 (2006)

75. T. N. Shorey, A. J. van der Poorten, R. Tijdeman and A. Schinzel, Applications of the Gel'fond-Baker method to Diophantine equations, In: Transcendence Theory: Advances and Applications, *Academic Press*, London-New York, San Francisco (1977), 59–77
76. T.N. Shorey, R. Tijdeman, *Exponential Diophantine equations, Cambridge Tracts in Mathematics 87* (Cambridge University Press, Cambridge, 1986)
77. C.L. Siegel, Approximation algebraischer Zahlen. Math. Zeit. **10**, 173–213 (1921)
78. S. Siksek, J.E. Cremona, On the Diophantine equation $x^2 + 7 = y^m$. Acta Arith. **109**, 143–149 (2003)
79. S. Siksek, On the Diophantine equation $x^2 = y^p + 2^k z^p$. J. Theor. Nombres Bordeaux **15**, 839–846 (2003)
80. G. Soydan, M. Demirci, I. Cangul, On the Diophantine equation $x^2 + 11^m = y^n$. Adv. Stud. Contemp. Math. (Kyungshang) **19**(2), 183–188 (2009)
81. G. Soydan, On the Diophantine equation $x^2 + 7^\alpha 11^\beta = y^n$. Miskolc Math. Notes **13**, 515–527 (2012)
82. G. Soydan, M. Ulas, H. Zhu, On the Diophantine equation $x^2 + 2^a 19^b = y^n$ Indian. J. Pure Appl. Math. **43**, 251–261 (2012)
83. G. Soydan, A note on the Diophantine equations $x^2 \pm 5^\alpha p^n = y^n$, *Commun. Fac. Sci. Univ. Ank. Sér. A1 Math. Stat.* **67** (2018), no. 1, 317–322
84. R. Steiner, N. Tzanakis, Simplifying the solution of Ljunggren's $x^2 + 1 = 2y^4$. J. Number Theory **37**, 123–132 (1991)
85. B. Stolt, Die Anzahl von Losungen gewisser diophantischer Gleichungen. Arch. Math. (Basel) **8**, 393–400 (1957)
86. B. Sury, On the Diophantine equation $x^2 + 2 = y^n$. Arch. Math. (Basel) **74**, 350–355 (2000)
87. Sz. Tengely ,On the Diophantine equation $x^2 + a^2 = 2y^p$, *Indag. Math. (NS)* **15** (2004), no. 2, 291–304
88. Sz. Tengely, On the Diophantine equation $x^2 + q^{2m} = 2y^p$,*Acta Arith.* **127** (2007), no. 1, 71–86
89. H.L. Zhu, A note on the Diophantine equation $x^2 + q^m = y^3$. Acta Arith. **146**, 195–202 (2011)
90. H. Zhu, M.H. Le, A. Togbe, On the exponential Diophantine equation $x^2 + p^{2m} = 2y^n$. Bull. Aust. Math. Soc. **86**, 303–314 (2012)

Partial Dedekind Zeta Values and Class Numbers of R–D Type Real Quadratic Fields

Mohit Mishra

2010 Mathematics Subject Classification Primary: 11R29 · 11R42 · Secondary: 11R11

1 Introduction

A number field k is a finite extension of the field of rational numbers $\mathbb{Q}$. Quadratic (number) field is a degree 2 extension of $\mathbb{Q}$. Every quadratic field k is of the form $\mathbb{Q}(\sqrt{d})$, where d is a square-free integer. If d is positive (respectively, negative), then the quadratic field $k = \mathbb{Q}(\sqrt{d})$ is called real (respectively, imaginary). A number $\beta \in \mathbb{C}$ is said to be an algebraic integer if β is a root of a nonzero, monic polynomial over $\mathbb{Z}$. The set of all algebraic integers in k forms a ring which is known as ring of integers of k. It is denoted by $\mathcal{O}_k$. In general, unlike in $\mathbb{Z}$, the unique factorization of algebraic integers into primes (irreducibles) does not hold in $\mathcal{O}_k$, that is, $\mathcal{O}_k$ is not a principal ideal domain in general. Therefore, it is interesting and of considerable importance to understand how far $\mathcal{O}_k$ fails to be a PID. As a result, to measure this failure, the concept of the ideal class group or simply class group of a number field appeared. Starting from the time of Gauss, this topic has been extensively explored by many authors and thus there exist a vast amount of research articles. Gauss conjectured that there are only nine imaginary quadratic fields with class number 1. Baker [2] and Stark [40] proved this conjecture independently. Heegner [21] had already proved this conjecture but unfortunately, his proof was incorrect or at the best, incomplete. Stark found that the gap in the proof is very minor and he filled that gap and completed the proof in [41]. Gauss also gives a list of imaginary

M. Mishra (✉)
Harish-Chandra Research Institute, HBNI, Chhatnag Road,
Jhunsi 211019, Allahabad, India
e-mail: mohitmishra@hri.res.in

K. Chakraborty et al. (eds.), *Class Groups of Number Fields and Related Topics*,
https://doi.org/10.1007/978-981-15-1514-9_15

quadratic fields with very low class numbers, and he believes that list to be complete. Baker and Stark independently in [3, 39], respectively, and jointly in [4], completely classified the list of imaginary quadratic fields with class number 2. Oesterlé [36] gives the analogous list of imaginary quadratic fields with class number 3. Finally, all the imaginary quadratic fields with class numbers up to 100 were classified by Watkins [42]. On the other hand, the divisibility of the class numbers of number fields is also very important for understanding the structure of the class groups of number fields. Interested readers can consult [14–17, 23, 25, 26, 29, 32, 38, 43] and the references therein, for more details on the divisibility properties of the class numbers of quadratic number fields.

In this survey article, we discuss some interesting results concerning the criteria for the class number of certain Richaud–Degert type real quadratic number fields to be 1 and 2. We also provide the outlines of the proof of some of these important results. Due to the versatility of these problems, this survey may miss out some interesting references and thus some interesting results as well. Therefore this article is never claimed to be a complete survey.

2 R–D Type Real Quadratic Fields and Some Conjectures

We begin this section with the following definition and notation.

Definition 2.1 Let $d = n^2 + r$ with $d \neq 5$ be a square-free positive integer satisfying

$$r \mid 4n \text{ and } n < r \leq n.$$

In this case, the field $k = \mathbb{Q}(\sqrt{d})$ is called real quadratic field of Richaud–Degert (in brief, R–D) type.

The following result gives the fundamental unit in R–D type real quadratic fields. This result was proved by Degert [20] in 1958.

Proposition 2.1 (Degert [20]) *Let $k = \mathbb{Q}(\sqrt{n^2 + r})$ be a real quadratic field of Richaud–Degert type. Then fundamental unit ε of k and its norm $N(\varepsilon)$ are given as follows:*

$$\varepsilon = \begin{cases} n + \sqrt{n^2 + r}, & N(\varepsilon) = -\,sgn\ r, \quad \text{if } |r| = 1, \\ \frac{n+\sqrt{n^2+r}}{2}, & N(\varepsilon) = -\,sgn\ r, \quad \text{if } |r| = 4, \\ \frac{n^2+r}{|r|} + \frac{2n}{|r|}\sqrt{n^2 + r}, & N(\varepsilon) = 1, \quad \text{if } |r| \neq 1, 4. \end{cases}$$

Let $k = \mathbb{Q}(\sqrt{d})$ be a real quadratic field with fundamental discriminant D and $h(d)$ denote the class number of k. In modern terms, Gauss conjectured the following:

(C1) $h(d) = 1$ for infinitely many d. Another form is "there exist infinitely many real quadratic fields of the form $\mathbb{Q}(\sqrt{p})$, $p \equiv 1 \pmod 4$ of class number 1".

This conjecture is still open. In connection to (C1), Chowla and Friedlander [18] stated the following conjecture:

(C2) If $D = m^2 + 1$ is a prime with $m > 26$, then the class number of $\mathbb{Q}(\sqrt{D})$ is greater than 1.

This conjecture concludes that there are exactly seven real quadratic fields of the form $\mathbb{Q}(\sqrt{m^2+1})$ whose class number is 1, and they are corresponding to

$$m \in \{1, 2, 4, 6, 10, 14, 26\}.$$

In 1988, Mollin and Williams [35] proved this conjecture under the assumption of generalized Riemann hypothesis (GRH). Chowla also posted a conjecture analogous to (C2) on a general family of real quadratic fields. More precisely, he gave the following conjecture.

(C3) Let D be a square-free integer of the form $D = 4m^2 + 1$ for some positive integer m. Then there exist exactly six real quadratic fields $\mathbb{Q}(\sqrt{D})$ whose class number is 1. These six fields are corresponding to $D \in \{5, 17, 37, 101, 197, 677\}$.

Yokoi [44] investigated this conjecture. He, however, stated one more conjecture on another family of real quadratic fields. Precisely, he stated the following:

(C4) Let D be a square-free integer of the form $D = m^2 + 4$ for some positive integer m. Then there exist exactly six real quadratic fields $\mathbb{Q}(\sqrt{D})$ of class number one. These fields are corresponding to $D \in \{5, 13, 29, 53, 173, 293\}$.

Kim et al. [28] proved that at least one of (C3) and (C4) is true. They also concluded that there are at most seven real quadratic fields $\mathbb{Q}(\sqrt{D})$ whose class number is 1 for the other case. Biró proved the conjectures (C3) and (C4) in [6, 7]. On the other hand, Hoque and Saikia [24] proved that there are no real quadratic fields of the form $\mathbb{Q}(\sqrt{9(8n^2+r)+2})$ whose class number is 1 when $n \geq 1$ and $r = 5, 7$. Hoque and Chakraborty [22] showed that if $d = n^2p^2 + 1$ with $p \equiv \pm 1 \pmod 8$ a prime and n an odd integer, then the class group of $\mathbb{Q}(\sqrt{d})$ is always nontrivial. Recently, Chakraborty and Hoque [11] proved that if d is a square-free part of $an^2 + 2$, where $a = 9, 196$ and n is an odd integer, then the class group of $\mathbb{Q}(\sqrt{d})$ is always nontrivial.

It is more interesting to investigate some conditions for a real quadratic field to have a given class number, say N. Applying algebraic method, Yokoi [44] proved that for a positive integer m, the class group of $\mathbb{Q}(\sqrt{4m^2+1})$ is trivial if and only if $m^2 - x(x+1)$, $1 \leq x \leq n-1$, is a prime. In [33], Lu obtained this result using the theory of continued fractions. Kobayashi [30] determined some strong conditions that this as well as some other families of real quadratic fields to be of class number 1. However, Byeon and Kim [8] established some necessary and sufficient conditions for the class number of Richaud–Degert type real quadratic fields to be 1. Analogously, they also obtained some necessary and sufficient conditions for the class number of Richaud–Degert type real quadratic fields to be 2. On the other hand, Mollin [34] obtained some analogous conditions for class number 2 using the theory of continued fractions and algebraic arguments. Along the same lines, some criteria for the class numbers of some R–D type real quadratic fields to be 3 were deduced in [13]. Recently, the author along with Chakraborty and Hoque [12] classified the order 4 class groups of certain real quadratic fields of R–D type using some group theoretic arguments.

3 Dedekind Zeta Values

In this section, we discuss two different ways, due to Siegel and Lang, respectively, of computing special values of zeta functions attached to a real quadratic field. Throughout this section, if not stated, k is a real quadratic field of Richaud–Degert (R–D) type, more precisely $k = \mathbb{Q}(\sqrt{d})$ with radicand $d = n^2 + r$ satisfying $r \mid 4n$ and $-n < r \leq n$.

The Dedekind zeta function of a number field k is defined by

$$\zeta_k(s) = \sum_{a} \frac{1}{\mathfrak{N}(a)^s}, \qquad s = \sigma + i\tau \text{ and } \sigma > 1,$$

where the sum is running over all the integral ideals of k. We can also express this zeta function as an Euler product:

$$\zeta_k(s) = \prod_{\wp} \left(1 - \frac{1}{\mathfrak{N}(\wp)^s}\right)^{-1},$$

where product runs over all the integral prime ideals.

Zagier [45] described the following formula by direct analytic methods when k is a real quadratic field, by specializing Siegel's formula [37] for $\zeta_k(1-2n)$ for general k. For $n = 1$, we have the following form (see [45]).

Theorem 3.1 *Let k be a real quadratic field with discriminant D. Then*

$$\zeta_k(-1) = \frac{1}{60} \sum_{\substack{|t| < \sqrt{D} \\ t^2 \equiv D \pmod 4}} \sigma\left(\frac{D - t^2}{4}\right),$$

where $\sigma(n)$ denotes the sum of divisors of n.

Lang gives another method of computing special values of $\zeta_k(s)$, whenever k is a real quadratic field. Let $\mathfrak{A}$ be an ideal class of a real quadratic field $k = \mathbb{Q}(\sqrt{d})$ with discriminant D. And let $\{r_1, r_2\}$ be an integral basis of an integral ideal $\mathfrak{a}$ in $\mathfrak{A}^{-1}$ with. We define

$$\delta(\mathfrak{a}) := r_1 r_2' - r_1' r_2,$$

where r_1' and r_2' are the conjugates of r_1 and r_2, respectively.

Let ε be the fundamental unit of k. Then we can find a matrix $M = \begin{bmatrix} a & b \\ c & d \end{bmatrix}$ with integer entries satisfying

$$\varepsilon \begin{bmatrix} r_1 \\ r_2 \end{bmatrix} = M \begin{bmatrix} r_1 \\ r_2 \end{bmatrix}, \tag{3.1}$$

since $\{\varepsilon r_1, \varepsilon r_2\}$ is also an integral basis of $\mathfrak{a}$. We can now state the result of Lang [31]:

Theorem 3.2 *By keeping the above notations, we have*

$$\begin{aligned}\zeta_k(-1, \mathfrak{A}) = & \frac{\text{sgn}\,\delta(\mathfrak{a})\, r_2 r_2'}{360 N(\mathfrak{a}) c^3} \{(a+d)^3 - 6(a+d)N(\varepsilon) - 240c^3(\text{sgn}\, c) \\ & \times S^3(a, c) + 180ac^3(\text{sgn}\, c)S^2(a, c) - 240c^3(\text{sgn}\, c)S^3(d, c) \\ & + 180dc^3(\text{sgn}\, c)S^2(d, c)\},\end{aligned}$$

where $S^i(-, -)$ denotes the generalized Dedekind sum as defined in [1] *and $N(\mathfrak{a})$ represents the norm of $\mathfrak{a}$.*

We note that Banerjee, Chakraborty, and Hoque obtained some important formulae for computing zeta values attached to both real as well as imaginary quadratic fields in [5]. However, these formulas are not helpful here.

To apply Theorem 3.2, one needs to determine the values of a, b, c, d and generalized Dedekind sum. The following result of Kim [27] determines the values of a, b, c, and d.

Lemma 3.1 *The entries of matrix M are given by*

$$a = Tr\left(\frac{r_1 r_2' \varepsilon}{\delta(\mathfrak{a})}\right), \quad b = Tr\left(\frac{r_1 r_1' \varepsilon'}{\delta(\mathfrak{a})}\right), \quad c = Tr\left(\frac{r_2 r_2' \varepsilon}{\delta(\mathfrak{a})}\right) \text{ and}$$
$$d = Tr\left(\frac{r_1 r_2' \varepsilon'}{\delta(\mathfrak{a})}\right).$$

Furthermore, $\det(M) = N(\varepsilon)$ *and* $bc \neq 0$.

Proof By (3.1) and taking its conjugate, we get

$$\begin{bmatrix} \varepsilon r_1 & \varepsilon' r_1' \\ \varepsilon r_2 & \varepsilon' r_2' \end{bmatrix} = M \begin{bmatrix} r_1 & r_1' \\ r_1 & r_1' \end{bmatrix}. \tag{3.2}$$

Also we can see that

$$\begin{bmatrix} r_1 & r_1' \\ r_2 & r_2' \end{bmatrix}^{-1} = \frac{1}{\delta(\mathfrak{a})} \begin{bmatrix} r_2' & -r_1' \\ -r_2 & r_1 \end{bmatrix}.$$

Hence by multiplying above matrix in (3.2), we get the desired result. □

Kim also derived the following expressions in [27] for special values of generalized Dedekind sum by using reciprocity law. These expressions are also required to compute the special value of zeta functions for ideal classes of respective real quadratic fields.

Lemma 3.2 *For any positive even integer m, we have*

(i) $S^3(m \pm 1, 2m) = \pm S^1(m+1, 2m) = \mp \frac{m^4-50m^2+4}{960m^3}$.
(ii) $S^3(m+1, 4m) = \frac{-m^4-180m^3+410m^2-4}{7680m^3}$.
(iii) $S^3(m-1, 4m) = \frac{m^4-180m^3-410m^2+4}{7680m^3}$.
(iv) $S^2(m-1, 2m) = S^2(m+1, 2m) = \frac{m^4+100m^2-6}{1440m^3}$.
(v) $S^2(m-1, 4m) = S^2(m+1, 4m) = \frac{m^4+820m^2-6}{11520m^3}$.

Lemma 3.3 *For any positive integer m, we have*

(i) $S^2(\pm 1, m) = \frac{m^4+10m^2-6}{180m^3}$.
(ii) $S^3(\pm 1, m) = \pm\frac{-m^4+5m^2-4}{120m^3}$.

4 Class Number Criteria

In this section, we calculate the value $\zeta_k(-1, \mathfrak{A})$ for an ideal class $\mathfrak{A}$ in k, and then equate these values with $\zeta_k(-1)$ to derive the results. Throughout this section, k is a Richaud–Degert (R–D) type real quadratic field. The following result appeared in [8].

Theorem 4.1 *Let $k = \mathbb{Q}(\sqrt{n^2+r})$ be a real quadratic field of R–D type with $|r| \neq 1, 4$. If $n^2 + r \equiv 2, 3 \pmod 4$, then $h(d) = 1$ if and only if*

$$\zeta_k(-1) = \frac{4n^3(r^2+1) + 2nr(3r^2+5r+3)}{180r^2}.$$

Proof Let $\mathcal{I}$ denote the principal ideal class. Since $|r| \neq 1, 4$, so that the fundamental unit in k is given by

$$\varepsilon = \frac{n^2+r}{|r|} + \frac{2n}{|r|}\sqrt{n^2+r}.$$

In this case, $r_1 = \sqrt{n^2+r}$ and $r_2 = 1$ forms an integral basis of $\mathcal{O}_k$. Let $\mathfrak{a} = \mathcal{O}_k = [r_1, r_2]$. Then by Lemma 3.1, one gets

$$\begin{bmatrix} a & b \\ c & d \end{bmatrix} = \begin{bmatrix} \frac{2n^2+r}{|r|} & \frac{2n(n^2+r)}{|r|} \\ \frac{2n}{|r|} & \frac{2n^2+r}{|r|} \end{bmatrix}.$$

It is easy to observe that

$$\frac{2n^2+r}{|r|} = n\frac{2n}{|r|} + sgn(r) \equiv sgn(r)\left(\frac{2n}{|r|}\right).$$

By using Lemma (3.3), we get

$$\begin{aligned}240c^3(\text{sgn } c)S^3(a,c) &= 240c^3S^3\left(\frac{2n^2+r}{|r|}, \frac{2n}{|r|}\right)\\ &= 240\times 4^3 S^3\left(sgn(r), \frac{2n}{|r|}\right)\\ &= -\frac{8n}{r^4}(4n^4-5n^2r^2+r^4).\end{aligned}$$

Similarly,

$$\begin{aligned}180ac^3(\text{sgn } c)S^2(a,c) &= 180ac^3S^2\left(\frac{2n^2+r}{|r|}, \frac{2n}{|r|}\right)\\ &= -\frac{2n}{r^5}(2n^2+r)(8n^4+20n^2r^2-3r^4).\end{aligned}$$

Also,

$$(a+d)^3 - 6(a+d)N(\varepsilon) = 8sgn(r)\frac{(2n^2+r)^3}{r^3} - 12sgn(r)\frac{2n^2+r}{r}.$$

Substituting the above values in Lang's formula, we get

$$\zeta_k(-1,\mathcal{I}) = \frac{4n^3(r^2+1)+2nr(3r^2+5r+3)}{180r^2}.$$

We know that

$$\zeta_k(-1) \geq \zeta_k(-1,\mathcal{I})$$

and equality holds if and only if $h(d)=1$. □

In a similar fashion, one can obtain similar results for other cases. Hence, we summarize the criteria for class number 1 as follows:

Theorem 4.2 *Let $k=\mathbb{Q}(\sqrt{n^2+r})$ be a real quadratic field of R–D type. Then $h(d)=1$, for each case, if and only if we have the following value of $\zeta_k(-1)$*

I. *If* $n^2+r\equiv 2 \pmod 4$ *and* $|r|=1$

$$\zeta_k(-1) = \frac{4n^3+5n\pm 6n}{180}$$

II. *If* $n^2+r\equiv 1 \pmod 4$

$$\zeta_k(-1) = \begin{cases} \frac{n^3+5n\pm 6n}{360}, & if|r| = 4, \\ \frac{n^3+14n}{360}, & if|r| = 1. \end{cases}$$

and if $|r| \neq 1, 4$

$$\zeta_k(-1) = \begin{cases} \frac{2n^3(r^2+1)+n(3r^3+50r^2+3r)}{720r^2}, & \text{if } n \text{ even}, \\ \frac{2n^3(r^2+16)+n(3r^3+20r^2+48r)}{720r^2}, & \text{if } n \text{ odd}. \end{cases}$$

One needs the following result to derive class number 2 criteria.

Theorem 4.3 *Let* $k = \mathbb{Q}(\sqrt{d})$ *be a real quadratic field with* d *be square-free. Then we have the following results:*

I. *If* $d \equiv 1 \pmod 8$, *then* (2) *splits, i.e.,*

$$(2) = \left(2, \frac{1+\sqrt{d}}{2}\right)\left(2, \frac{1-\sqrt{d}}{2}\right).$$

II. *If* $d \equiv 2, 3 \pmod 4$, *then* (2) *ramifies, i.e.,*

$$(2) = \begin{cases} (2, d)^2, & \text{if } d \equiv 2 \pmod 4, \\ (2, 1+d)^2, & \text{if } d \equiv 3 \pmod 4. \end{cases}$$

III. *If* $d \equiv 5 \pmod 8$, *then* (2) *remains prime.*

One can consult [19] for detail proof of this result.

Let $\mathfrak{A}$ be the ideal class containing $\left(2, \frac{1\pm\sqrt{d}}{2}\right)$ or $(2, \alpha + d)$, where $\alpha = 0$ or 1 depending on $d \equiv 2 \pmod 4$ or $d \equiv 2 \pmod 4$. Now, one can get the following result.

Theorem 4.4 *Let* $k = \mathbb{Q}(\sqrt{n^2+r})$ *be a real quadratic field of R–D type. Then the following hold:*

I. *If* $n^2 + r \equiv 2 \pmod 4$, *then*

$$\zeta_k(-1, \mathfrak{A}) = \begin{cases} \frac{2n^3(r^2+1)+nr(3r^2+50r+3)}{360r^2}, & \text{if } n \text{ odd and } |r| \neq 1, 4, \\ \frac{2n^3(r^2+16)+nr(3r^2+20r+48)}{360r^2}, & \text{if } n \text{ even and } |r| \neq 1, 4, \\ \frac{2n^3+25n\pm 3n}{360}, & \text{if } |r| = 1. \end{cases}$$

II. *If* $n^2 + r \equiv 3 \pmod 4$, *then*

$$\zeta_k(-1, \mathfrak{A}) = \begin{cases} \frac{2n^3(r^2+1)+nr(3r^2+50r+3)}{360r^2}, & \text{if } n \text{ even and } |r| \neq 1, 4, \\ \frac{2n^3(r^2+16)+nr(3r^2+20r+48)}{360r^2}, & \text{if } n \text{ odd and } |r| \neq 1, 4, \\ \frac{2n^3+25n\pm 3n}{360}, & if|r| = 1. \end{cases}$$

III. *If* $n^2 + r \equiv 1 \pmod 8$, *then*

$$\zeta_k(-1, \mathfrak{A}) = \begin{cases} \frac{2n^3(r^2+1)+n(3r^3+410r^2+3r)}{2880r^2}, & if |r| \neq 1, 4, \\ \frac{n^3+104n}{1440}, & if |r| = 1. \end{cases}$$

Proof We will give the proof for the case when $n^2 + r \equiv 1 \pmod 8$ and $|r| = 1$. Let $\mathfrak{a} := \left(2, \frac{1+\sqrt{d}}{2}\right) \in \mathfrak{A}^{-1}$. Then an integral basis for $\mathfrak{a}$ is $\{r_1 = \frac{1+\sqrt{d}}{2},\ r_2 = 2\}$ and thus $\delta(\mathfrak{a}) = 2\sqrt{d}$. By Lemma 3.1, we get

$$\begin{bmatrix} a & b \\ c & d \end{bmatrix} = \begin{bmatrix} n+1 & \frac{d-1}{4} \\ 4 & n-1 \end{bmatrix}.$$

Since $n^2 + 1 \equiv 1 \pmod 8$, so $4|n$, and therefore $n \pm 1 \equiv \pm 1 \pmod 4$. Hence by Lemma 3.3, we obtain

$$240c^3(sgn\,)S^3(d, c) = 240c^3 S^3(n-1, 4) = 240 \times 4^3 S^3(-1, 4) = 360,$$

$$240c^3(sgn\, c)S^3(a, c) = 240c^3 S^3(n+1, 4) = 240 \times 4^3 S^3(1, 4) = -360,$$

$$180dc^3(sgn\,)S^2(d, c) = 180dc^3 S^2(n-1, 4) = 180 \times 4^3 d S^2(-1, 4) = 410(n-1).$$

$$180ac^3(sgn\,)S^2(a, c) = 180ac^3 S^2(n+1, 4) = 180 \times 4^3 a S^2(1, 4) = 410(n+1),$$

Therefore by Theorem 3.2, we have

$$\zeta_k(-1, \mathfrak{A}) = \frac{n^3 + 104n}{1440}.$$ □

Theorem 4.5 *Let* $k = \mathbb{Q}(\sqrt{n^2 + r})$ *be R–D type real quadratic field. Then*

I. $d = n^2 + r \equiv 1 \pmod 8$
(i) $h(d) > 1$ *for* $|r| = 1$ *except* $d = 17$.
(ii) $h(d) > 1$ *for* $|r| \neq 1, 4$ *except* $d = 33$.

II. $d = n^2 + r \equiv 2, 3 \pmod 4$
(i) $h(d) > 1$ *for* $|r| = 1$ *except* $d = 2, 3$.
(ii) $h(d) > 1$ *for* $|r| \neq 1, 4$ *except* $r = \pm 2$.

Proof We will give the details of the proof for the case **I**(i), and other cases can be handled along the same lines.

Let $n^2 + 1 \equiv 1 \pmod 8$. Then by above theorem,

$$\zeta_k(-1, \mathfrak{A}) = \frac{n^3 + 104n}{1440}.$$

and by Theorem 4.2,

$$\zeta_k(-1,\mathcal{I}) = \frac{n^3+14n}{360}.$$

If $h(d)=1$, then $\zeta_k(-1,\mathcal{I}) = \zeta_k(-1,\mathfrak{A})$, i.e.,

$$\frac{n^3+104n}{1440} = \frac{n^3+14n}{360}.$$

Thus we get $d=17$. Hence for $d\neq 17$, we have $h(d)>1$.

Theorem 4.6 *Let* $k=\mathbb{Q}(\sqrt{n^2+r})$ *be a real quadratic field of R–D type. Then* $h(d)=2$ *if and only if*

I. *If* $d=n^2+r\equiv 2 \pmod 4$

$$\zeta_k(-1) = \begin{cases} \frac{2n^3(r^2+1)+nr(3r^3+14r^2+3r)}{72r^2}, & \text{if } n \text{ odd, } |r|\neq 2 \text{ and } |r|\neq 1,4,\\ \frac{2n^3(r^2+4)+n(3r^3+8r^2+12r)}{72r}, & \text{if } n \text{ even } |r|\neq 2 \text{ and } |r|\neq 1,4,\\ \frac{10n^3+35n\pm 15n}{360}, & \text{if } d\neq 2,3 \text{ and } |r|=1.\end{cases}$$

II. *If* $d=n^2+r\equiv 3 \pmod 4$

$$\zeta_k(-1,\mathfrak{A}) = \begin{cases} \frac{2n^3(r^2+1)+nr(3r^3+14r+3r)}{72r^2}, & \text{if } n \text{ even, } |r|\neq 2 \text{ and } |r|\neq 1,4,\\ \frac{2n^3(r^2+4)+n(3r^3+8r^2+12r)}{72r}, & \text{if } n \text{ odd } |r|\neq 2 \text{ and } |r|\neq 1,4,\\ \frac{10n^3+35n\pm 15n}{360}, & \text{if } d\neq 2,3 \text{ and } |r|=1.\end{cases}$$

III. *If* $d=n^2+r\equiv 1 \pmod 8$

$$\zeta_k(-1) = \begin{cases} \frac{2n^3(r^2+1)+n(3r^3+122r^2+3r)}{576r^2}, & \text{if } n \text{ even, } d\neq 33 \text{ and } |r|\neq 1,4,\\ \frac{2n^3(r^2+13)+n(3r^3+98r^2+39r)}{576r^2}, & \text{if } n \text{ odd, } d\neq 33 \text{ and } |r|\neq 1,4,\\ \frac{n^3+32n}{288}, & \text{if } d\neq 17 \text{ and } |r|=1.\end{cases}$$

We note that detailed proof of this theorem can be found in [9].

Acknowledgement The author is thankful to the anonymous referee(s) for his/her valuable comments which have helped improving the readability of this manuscript

References

1. T.M. Apostol, Generalized Dedekind sums and transformation formulae of certain Lambert series. Duke Math. J. **17**, 147–157 (1950)
2. A. Baker, Linear forms in the logarithms of algebraic numbers I, II, III. Mathematika **13**, 204–216 (1966); **14**, 102–107 (1967); **14**, 220–228 (1967)

3. A. Baker, Imaginary quadratic fields with class number 2. Ann. Math. **94**(2), 139–152 (1971)
4. A. Baker, H. Stark, On a fundamental inequality in number theory. Ann. Math. **94**(2), 190–199 (1971)
5. S. Banerjee, K. Chakraborty, A. Hoque, *An analogue of Wilton's formula and values of Dedekind zeta functions* (2019), arXiv:1611.08693
6. A. Biró, Yokoi's conjecture. Acta Arith. **106**, 85–104 (2003)
7. A. Biró, Chowla's conjecture. Acta Arith. **107**, 179–194 (2003)
8. D. Byeon, H.K. Kim, Class number 1 criteria for real quadratic fields of Richaud–Degert type. J. Number Theory **57**, 328–339 (1996)
9. D. Byeon, H.K. Kim, Class number 2 criteria for real quadratic fields of Richaud–Degert type. J. Number Theory **62**, 257–272 (1998)
10. D. Byeon, J. Lee, Class number 2 problem for certain real quadratic fields of Richaud–Degert type. J. Number Theory **128**, 865–883 (2008)
11. K. Chakraborty, A. Hoque, *On the plus parts of the class numbers of cyclotomic fields*. Preprint
12. K. Chakraborty, A. Hoque, M. Mishra, A classification of order 4 class groups of $Q(\sqrt{n^2+1})$. Under review, arxiv:1902.05250
13. K. Chakraborty, A. Hoque, M. Mishra, A note on certain real quadratic fields with class number up to three. Kyushu J. Math. (to appear), arXiv:1812.02488
14. K. Chakraborty, A. Hoque, Class groups of imaginary quadratic fields of 3-rank at least 2. Ann. Univ. Sci. Budapest. Sect. Comput. **47**, 179–183 (2018)
15. K. Chakraborty, A. Hoque, Divisibility of class numbers of certain families of quadratic fields. J. Ramanujan Math. Soc. **34**(3), 281–289 (2019)
16. K. Chakraborty, A. Hoque, Y. Kishi, P.P. Pandey, Divisibility of the class numbers of imaginary quadratic fields. J. Number Theory **185**, 339–348 (2018)
17. K. Chakraborty, A. Hoque, R. Sharma, Divisibility of class numbers of quadratic fields: qualitative aspects, *Advances in Mathematical inequalities and Application*, Trends in Mathematics (Birkhäuser/Springer, Singapore, 2018), pp. 247–264
18. S. Chowla, J. Friedlander, Class numbers and quadratic residues. Glasg. Math. J. **17**(1), 47–52 (1976)
19. H. Cohn, *Advanced Number Theory* (Dover, New York, 1961)
20. G. Degert, Über die Bestimmung der Grundeinheit gewisser reell-quadratischer Zhalkörper. Abh. Math. Sem. Univ. Hamburg. **22**, 92–97 (1958)
21. K. Heegner, Diophantische analysis und modulfunktionen. Math Z. **56**, 227–253 (1952)
22. A. Hoque, K. Chakraborty, Pell-type equations and class number of the maximal real subfield of a cyclotomic field. Ramanujan J. **46**(3), 727–742 (2018)
23. A. Hoque, H.K. Saikia, On generalized Mersenne Primes and class-numbers of equivalent quadratic fields and cyclotomic fields. SeMA J. **67**(1), 71–75 (2015)
24. A. Hoque, H.K. Saikia, On the class-number of the maximal real subfield of a cyclotomic field. Quaest. Math. **37**(7), 889–894 (2016)
25. A. Hoque, H.K. Saikia, A note on quadratic fields whose class numbers are divisible by 3. SeMA J. **73**(1), 1–5 (2016)
26. A. Ito, A note on the divisibility of class numbers of imaginary quadratic fields $\mathbb{Q}(\sqrt{a^2-k^n})$. Proc. Japan Acad. Ser. A **87**, 151–155 (2011)
27. H.K. Kim, A conjecture of S. Chowla and related topics in analytic number theory. Ph.D. thesis, The Johns Hopkins University (1988)
28. H.K. Kim, M.-G. Leu, T. Ono, On two conjectures on real quadratic fields. Proc. Jpn. Acad. Ser. A Math. Sci. **63**(6), 222–224 (1987)
29. Y. Kishi, Note on the divisibility of the class number of certain imaginary quadratic fields. Glasg. Math. J. **51**, 187–191 (2009); corrigendum, ibid. **52**, 207–208 (2010)
30. M. Kobayashi, Prime producing quadratic polynomials and class number 1 problem for real quadratic fields. Proc. Jpn. Acad. Ser. A Math. Sci. **66**(5), 119–121 (1990)
31. H. Lang, Über eine Gattung elemetar-arithmetischer Klassen invarianten reell-quadratischer Zhalkörper. J. Reine Angew. Math. **233**, 123–175 (1968)

32. S.R. Louboutin, On the divisibility of the class number of imaginary quadratic number fields. Proc. Am. Math. Soc. **137**, 4025–4028 (2009)
33. H. Lu, On the real quadratic fields of class-number one. Sci. Sin. **24**(10), 1352–1357 (1981)
34. R.A. Mollin, Applications of a new class number two criterion for real quadratic fields. *Computational Number Theory* (Debrecen, 1989) (de Gruyter, Berlin, 1991), pp. 83–94
35. R.A. Mollin, H.C. Williams, A conjecture of S. Chowla via the generalized Riemann hypothesis. Proc. Am. Math. Soc. **102**, 794–796 (1988)
36. J. Oesterlé, Le problme de Gauss sur le nombre de classes (French). Enseign. Math. **34**(1–2), 43–67 (1988)
37. C.L. Siegel, Berechnung von Zetafunktionen an ganzzahligen Stellen. Nachr. Akad. Wiss. Göttingen Math.-Phys. Kl. II **10**, 87–102 (1969)
38. K. Soundararajan, Divisibility of class numbers of imaginary quadratic fields. J. Lond. Math. Soc. **61**, 681–690 (2000)
39. H. Stark, A transcendence theorem for class-number problems I, II. Ann. Math. **94**(2), 153–173 (1971); **96**, 174–209 (1972)
40. H. Stark, A complete determination of the complex quadratic fields of class number one. Mich. Math. J. **14**, 1–27 (1967)
41. H. Stark, On the "Gap" in a theorem of Heegner. J. Number Theory **1**, 16–27 (1969)
42. M. Watkins, Class numbers of imaginary quadratic fields. Math. Comp. **73**(246), 907–938 (2004)
43. Y. Yamamoto, On unramified Galois extensions of quadratic number fields. Osaka J. Math. **7**, 57–76 (1970)
44. H. Yokoi, Class-number one problem for certain kind of real quadratic fields, in *Proceedings of International Conference on Class Numbers and Fundamental Units of Algebraic Number Fields* (Katata, 1986) (Nagoya University, Nagoya, 1986), pp. 125–137
45. D. Zagier, On the values at negative integers of the zeta function of a real quadratic fields. Enseign. Math. **19**, 55–95 (1976)

On the Continued Fraction Expansions of $\sqrt{p}$ and $\sqrt{2p}$ for Primes $p \equiv 3 \pmod 4$

Stéphane R. Louboutin

2010 Mathematics Subject Classification Primary. 11A55 · 11R11

1 Introduction

Looking for example at the following continued fraction expansions of $\sqrt{d}$ for $d = p$ and $d = 2p$, where p is a prime integer equal to 3 modulo 4, one is lead to guess the behavior given in Theorem 1:

$d = p$ or $d = 2p$	$p \mod 8$	L	a_0	a_L	$\sqrt{d}$
$d = p = 43$	3	5	6	5	$[6, \overline{1, 1, 3, 1, 5, 1, 3, 1, 1, 12}]$
$d = 2p = 2 \cdot 43$	3	5	9	8	$[9, \overline{3, 1, 1, 1, 8, 1, 1, 1, 3, 18}]$
$d = p = 59$	3	3	7	7	$[7, \overline{1, 2, 7, 2, 1, 14}]$
$d = 2p = 2 \cdot 59$	3	5	10	10	$[10, \overline{1, 6, 3, 2, 10, 2, 3, 6, 20}]$
$d = p = 31$	7	4	5	5	$[5, \overline{1, 1, 3, 5, 3, 1, 1, 10}]$
$d = 2p = 2 \cdot 31$	7	2	7	6	$[7, \overline{1, 6, 1, 14}]$
$d = p = 47$	7	2	6	5	$[6, \overline{1, 5, 1, 12}]$
$d = 2p = 2 \cdot 47$	7	8	9	8	$[9, \overline{1, 2, 3, 1, 1, 5, 1, 8, 1, 5, 1, 1, 3, 2, 1, 18}]$

This behavior in the case of $d = p$ was presented orally in [8] and in written form in [1, Proposition 4.1]. However it had already been proved in [2, Corollary 2, p. 2071]. Our present proof is different and applies both to $d = p$ and $d = 2p$. It is based on the arithmetic of quadratic number fields and their ideal class groups in the narrow sense (as in [5, 6]).

S. R. Louboutin (✉)
Aix Marseille Université, CNRS, Centrale Marseille, I2M, Marseille, France
e-mail: stephane.louboutin@univ-amu.fr

K. Chakraborty et al. (eds.), *Class Groups of Number Fields and Related Topics*,
https://doi.org/10.1007/978-981-15-1514-9_16

Theorem 1 *Let $p \equiv 3 \pmod 4$ be a prime integer. Let $l \geq 1$ be the length of the period of the continued fraction expansion $\sqrt{d} = [a_0, \overline{a_1, \ldots, a_l}]$ of $d = p$ or $d = 2p$. Then (i) $a_0 = \lfloor \sqrt{d} \rfloor$ and $a_l = 2a_0$, (ii) $a_k = a_{l-k}$ for $1 \leq k \leq l-1$, (iii) $l = 2L$ is even, (iv) $a_{l/2} = a_L$ is the integer in $\{a_0 - 1, a_0\}$ of the same parity as d. Moreover, L is even if and only if $p \equiv 7 \pmod 8$.*

2 Continued Fraction Expansions of Quadratic Irrationalities

Let $d > 1$ be a not perfect square integer. The *continued fraction expansion* $\omega_0 = [a_0, a_1, \ldots]$ of $\omega_0 = (P_0 + \sqrt{d})/Q_0$ with $P_0, Q_0 \in \mathbb{Z}$, and Q_0 dividing $d - P_0^2$, can be computed inductively by writing $\omega_k = [a_k, \ldots]$ as $\omega_k = (P_k + \sqrt{d})/Q_k$, where the $P_k, Q_k \in \mathbb{Z}$ are inductively defined by $a_k = \lfloor (P_k + \sqrt{d})/Q_k \rfloor$ and $\omega_k = a_k + 1/\omega_{k+1}$, hence by $P_{k+1} = a_k Q_k - P_k$ and $Q_{k+1} = (d - P_{k+1}^2)/Q_k$. (Hence $Q_{k+1} = Q_{k-1} + 2a_k P_k - a_k^2 Q_k$ for $k \geq 1$ and the Q_k's are rational integers, by induction on k). Recall that $\omega_0 \in \mathbb{Q}(\sqrt{d})$ is called *reduced* if $\omega_0 > 1$ and $-1/\omega_0' > 1$, where ω_0' is the conjugate of ω_0 in $\mathbb{Q}(\sqrt{d})$. By induction, using $\omega_k = a_k + 1/\omega_{k+1}$, it is easy to see that if ω_0 is reduced then so are all the ω_k's. The continued fraction expansion $\alpha = [\overline{a_1, \ldots, a_l}]$ of an irrational real number α is *purely periodic* (of *length* l) if and only if α is a reduced quadratic irrationality of the form $\alpha = (P + \sqrt{d})/Q$ for some not perfect square integer $d > 1$ and some $P \in \mathbb{Z}$, $Q \in \mathbb{Z}_{\geq 1}$ and Q dividing $d - P^2$. In that case, $-1/\alpha' = [\overline{a_l, \ldots, a_1}]$ (e.g., see [3, XV, p. 311]). Now, if $\omega_0 = [\overline{a_0, a_1, \ldots, a_{l-1}}]$ is reduced, using $\omega_k = a_k + 1/\omega_{k+1}$, we obtain $\mathbb{M}_k := \mathbb{Z} + \mathbb{Z}\omega_k = \mathbb{Z} + \mathbb{Z}\omega_{k+1}^{-1} = \omega_{k+1}^{-1}\mathbb{M}_{k+1}$ and $\mathbb{M}_0 = \omega_1^{-1}\mathbb{M}_1 = \omega_1^{-1}\omega_2^{-1}\mathbb{M}_2 = \cdots = \varepsilon^{-1}\mathbb{M}_l = \varepsilon^{-1}\mathbb{M}_0$, where $\varepsilon = \omega_1\omega_2 \ldots \omega_l = \omega_0\omega_1 \ldots \omega_{l-1}$. Therefore, ε is a unit of the module $\mathbb{M}_0 = \mathbb{Z} + \mathbb{Z}\omega_0 \subseteq \mathbb{Q}(\sqrt{d})$. Hence ε is a unit of norm $N(\varepsilon) = \prod_{k=0}^{l-1}(\omega_k\omega_k') = (-1)^l$ (as $\omega_k > 1$ and $-1/\omega_k' > 1$).

Now, setting $g = \lfloor \sqrt{d} \rfloor$, it is easy to check that $\omega_0 = g + \sqrt{d}$ is reduced. Its continued fraction expansion $\omega_0 = [\overline{2g, a_1, \ldots, a_{l-1}}]$ is therefore purely periodic and $\omega_1 = [\overline{a_1, \ldots, a_{l-1}, 2g}] = 1/(\omega_0 - 2g) = 1/(\sqrt{d} - g) = -1/\omega_0' = [\overline{a_{l-1}, \ldots, a_1, 2g}]$. Hence, $a_k = a_{l-k}$ for $1 \leq k \leq l-1$. Assume that d is divisible by a prime $p \equiv 3 \pmod 4$. The unit $\varepsilon = \omega_0\omega_1 \ldots \omega_{l-1} = x_d + y_d\sqrt{d} \in \mathbb{M}_0 := \mathbb{Z} + \mathbb{Z}\omega_0 = \mathbb{Z}[\sqrt{d}]$ satisfies $(-1)^l = N(\varepsilon) = x_d^2 - dy_d^2 \equiv x_d^2 \pmod p$. Hence, $l = 2L$ is even, $\omega_0 = [\overline{2g, a_1, \ldots, a_{L-1}, a_L, a_{L-1}, \ldots, a_1}]$, $\omega_L = [\overline{a_L, \ldots, a_1, 2g, a_1, \ldots, a_{L-1}}]$ and $\omega_{L+1} = [\overline{a_{L-1}, \ldots, a_1, 2g, a_1, \ldots, a_L}] = -1/\omega_L'$. Using $\omega_{L+1} = (P_{L+1} + \sqrt{d})/Q_{L+1}$ and $\omega_L = (P_L + \sqrt{d})/Q_L$, we obtain $Q_L Q_{L+1} = (P_{L+1} + \sqrt{d})(\sqrt{d} - P_L)$. Looking at the coefficients of $\sqrt{d}$ in this identity, we obtain $P_{L+1} = P_L$. Hence, $P_L = P_{L+1} = a_L Q_L - P_L$ and $d - P_L^2 = Q_L Q_{L+1}$, i.e.,

$$2P_L = a_L Q_L, \text{ and } 4d - a_L^2 Q_L^2 = 4Q_L Q_{L+1}.$$

Hence, Q_L divides $4d$ and if 4 divides Q_L, then 4 divides d. Hence, we obtain

Lemma 2 *Let $d \equiv 2, 3 \pmod 4$ be a positive square-free integer such that at least one prime $p \equiv 3 \pmod 4$ divides d. Hence $\mathbb{Z}[\sqrt{d}]$ is the ring of algebraic integers of the real quadratic field $\mathbb{Q}(\sqrt{d})$ of discriminant $4d$ and the units of $\mathbb{Z}[\sqrt{d}]$ are of norm $+1$. Let $l \geq 1$ be the length of the period of the continued fraction expansion $\sqrt{d} = [a_0, \overline{a_1, \ldots, a_l}]$ of $d = p$ or $d = 2p$. Then (i) $a_0 = \lfloor \sqrt{d} \rfloor$ and $a_l = 2a_0$, (ii) $a_k = a_{l-k}$ for $1 \leq k \leq l-1$, (iii) $l = 2L$ is even, and (iv) Q_L is a square-free integer dividing $4d$ such that $1 < Q_L < 2\sqrt{d}$ and $a_L = \lfloor \omega_L \rfloor \in \{\lfloor 2\sqrt{d}/Q_L \rfloor, \lfloor 2\sqrt{d}/Q_L \rfloor - 1\}$. Moreover, L is even if and only if the ideal $\mathcal{I} = Q_L\mathbb{Z} + (P_L + \sqrt{d})\mathbb{Z}$ of norm Q_L which is principal in the wide sense is also principal in the narrow sense.*

Proof Set $\alpha = Q_L\omega_1 \ldots \omega_L \in \mathbb{Q}(\sqrt{d})$. Since $\mathcal{I} = Q_L\mathbb{M}_L = Q_L\omega_1 \ldots \omega_L\mathbb{M}_0 = \alpha\mathbb{Z}[\sqrt{d}]$, this ideal $\mathcal{I}$ is principal. Since the sign of the norm of α is $(-1)^L$ (recall that $\omega_k > 1$ and $-1/\omega_k' > 1$ for $k \geq 0$), the ideal $\mathcal{I}$ is principal in the narrow sense if and only if L is even. Finally, $\omega_L = (P_L + \sqrt{d})/Q_L$ is reduced if and only if $Q_L \geq 1$ and $|\sqrt{d} - Q_L| < P_L < \sqrt{d}$, which implies $1 < Q_L < 2\sqrt{d}$ and $\frac{2\sqrt{d}}{Q_L} - 1 < \omega_L < \frac{2\sqrt{d}}{Q_L}$. □

Notice that *(iii)* is related to [7, Satz 14, p. 94] and [4, Theorem 1].

3 Proof of Theorem 1

Lemma 3 *Let $p \equiv 3 \pmod 4$ be a prime integer. Take $d = p$ or $d = 2p$ and let the notation be as in Lemma 2. Then $Q_L = 2$. Hence, $\mathcal{I}$ is the prime ramified ideal $\mathcal{P}_2$ of norm 2 of the ring of algebraic integers $\mathbb{Z}[\sqrt{d}]$ of the real quadratic field $\mathbb{Q}(\sqrt{d})$ of discriminant $4d$ and a_L is the integer in $\{\lfloor \sqrt{d} \rfloor, \lfloor \sqrt{d} \rfloor - 1\}$ of the same parity as d.*

Proof Assume that $d = p$ or $d = 2p$, where $7 < p \equiv 3 \pmod 4$ is prime. Then Q_L is square-free, $1 < Q_L < 2\sqrt{d} \leq 2\sqrt{2p} < p$ and Q_L divides $4d = 4p$ or $8p$. Hence $Q_L = 2$. For $p = 3$, we have $\sqrt{3} = [1, \overline{1, 2}]$ and $\sqrt{6} = [2, \overline{2, 4}]$, and for $p = 7$, we have $\sqrt{7} = [2, \overline{1, 1, 1, 4}]$ and $\sqrt{14} = [3, \overline{1, 2, 1, 6}]$ and in these four cases, we have $Q_L = 2$. Since $Q_L = 2$, we have $a_L \in \{\lfloor \sqrt{d} \rfloor, \lfloor \sqrt{d} \rfloor - 1\}$ and $4d - a_L^2 Q_L^2 = 4p - 4a_L^2 = 4Q_L Q_{L+1} = 8Q_{L+1}$ implies that a_L has the same parity as d. □

Lemma 4 *Let $p \equiv 3 \pmod 8$ be a prime integer. Take $d = p$ or $d = 2p$. The prime ideal $\mathcal{P}_2$ of norm 2 of the ring of algebraic integers $\mathbb{Z}[\sqrt{d}]$ of the real quadratic field $\mathbb{Q}(\sqrt{d})$ of discriminant $4d$ is principal in the narrow sense if and only if $p \equiv 7 \pmod 8$.*

Proof Since the discriminant $4d$ of $\mathbb{Q}(\sqrt{d})$ has exactly 2 distinct prime divisors, the 2-rank of its narrow ideal class group is equal to 1 and the class of order 2 in the narrow class group is either the class of the prime ideal $\mathcal{P}_2$ above the prime 2 or the class of the prime ideal $\mathcal{P}_p$ above the prime p.

First, assume that $d = p$. If $\mathcal{P}_p = (\alpha) = (x + y\sqrt{p})$ were principal in the narrow class group, then we would have $p = N(\alpha) = x^2 - py^2$, hence p would divide $x = pX$ and $1 = pX^2 - y^2$ would imply $1 \equiv -y^2 \pmod{p}$ and $p \equiv 1 \pmod{4}$, a contradiction. Hence, the ideal class of $\mathcal{P}_p$ is the class of order 2 in the narrow class group.

If $\mathcal{P}_2 = (\alpha) = (x + y\sqrt{p})$ is principal in the narrow sense, then $N(\mathcal{P}_2) = 2 = N(\alpha) = x^2 - py^2$. Hence, x and y are odd and $2 = x^2 - py^2 \equiv 1 - p \pmod{8}$, i.e., $p \equiv 7 \pmod{8}$.

If $\mathcal{P}_2$ is not principal in the narrow sense, then $\mathcal{P}_2\mathcal{P}_p = (\alpha) = (x + y\sqrt{p})$ is principal in the narrow sense and $N(\mathcal{P}_2\mathcal{P}_p) = 2p = N(\alpha) = x^2 - py^2$. Hence p divides $x = pX$ and $y = pY$ and $2 = pX^2 - Y^2$. Hence, X and Y are odd and $2 = pX^2 - Y^2 \equiv p - 1 \pmod{8}$, i.e., $p \equiv 3 \pmod{8}$.

Second, assume that $d = 2p$. If $\mathcal{P}_2\mathcal{P}_p = (\alpha) = (x + y\sqrt{2p})$ were principal in the narrow class group, then we would have $2p = N(\alpha) = x^2 - 2py^2$. Hence $2p$ would divide $x = 2pX$ and we would have$1 = 2pX^2 - y^2$. Hence, y would be odd and we would have $1 + y^2 \equiv 2 \pmod{8}$. Hence X would be odd and we would have $2p \equiv 2pX^2 \equiv 1 + y^2 \equiv 2 \pmod{8}$ and $p \equiv 1 \pmod{4}$, a contradiction. Hence, the ideal class of $\mathcal{P} - 2\mathcal{P}_p$ is the class of order 2 in the narrow class group.

If $\mathcal{P}_2 = (\alpha) = (x + y\sqrt{2p})$ is principal in the narrow sense, then $x^2 - 2py^2 = 2$. Hence y is odd, $x = 2X$ is even, $2X^2 - py^2 = 1$, and $p \equiv py^2 \equiv 2X^2 - 1 \equiv 1, 7 \pmod{8}$, i.e., $p \equiv 7 \pmod{8}$.

If $\mathcal{P}_2$ is not principal in the narrow sense, then $\mathcal{P}_p = (x + y\sqrt{2p})$ is principal in the narrow sense and $x^2 - 2py^2 = p$. Hence p divides $x = pX$ and $pX^2 - 2y^2 = 1$. Hence X is odd and $p \equiv pX^2 \equiv 1 + 2y^2 \equiv 1, 3 \pmod{8}$, i.e. $p \equiv 3 \pmod{8}$. □

Acknowledgements We thank Yasuhiro Kishi for pointing us Refs. [2, 7].

References

1. D. Chakraborty, A. Saikia, Explicit solutions to $u^2 - pv^2 = \pm 2$ and a conjecture of Mordell, 14 October 2018, arXiv:1810.05980v1
2. E.P. Golubeva, Quadratic irrationals with fixed period length in the continued fraction expansion. J. Math. Sci. **70**, 2059–2076 (1994)
3. H. Hasse, Vorlesungen über Zahlentheorie. *Zweite neubearbeitete Auflage. Die Grundlehren der Mathematischen Wissen schaften, Band*, vol. 59 (Springer, Berlin, 1964)
4. F. Kawamoto, Y. Kishi, K. Tomita, Continued fraction expansions with even period and primary symmetric parts with extremely large end. Comment. Math. Univ. St. Pauli **64**(2), 131–155 (2015)
5. S. Louboutin, Continued fractions and real quadratic fields. J. Number Theory **30**, 167–176 (1988)
6. S. Louboutin, Groupes des classes d'idéaux triviaux. Acta Arith. **54**, 61–74 (1989)
7. Perron, Die Lehre von den Kettenbrüchen. Bd I. Elementare Kettenbrüche. 3te Aufl (B. G. Teubner Verlagsgesellschaft, Stuttgart, 1954)
8. A. Saikia, On the continued fraction of $\sqrt{p}$, in *International Conference on Class Groups of Number Fields and Related Topics*, Harish-Chandra Research Institute, Allahabad, Inde, 9 October 2018

Printed in the United States
By Bookmasters